Learn more on Discord

Join over 11,000 members on their Machine Learning journey:

https://byhand.ai/discord

ISBN-13: 978-1-80667-477-0

First edition, May 2025

Name ____________________

Chapter	Date Completed
1. Dot Product	
2. Matrix Multiplication	
3. Linear Layer	
4. Activation	
5. Artificial Neuron	
6. Batch	
7. Connection	
8. Hidden Layer	
9. Deep	
10. Wide	
11. Softmax	
12. Gradient	

Table of Contents

Chapter 1

Dot Product

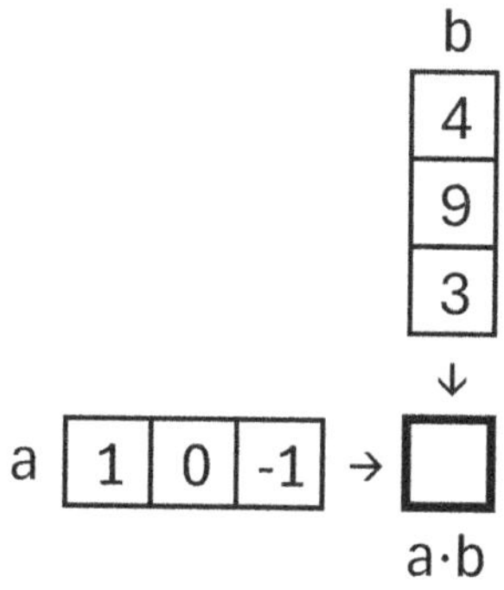

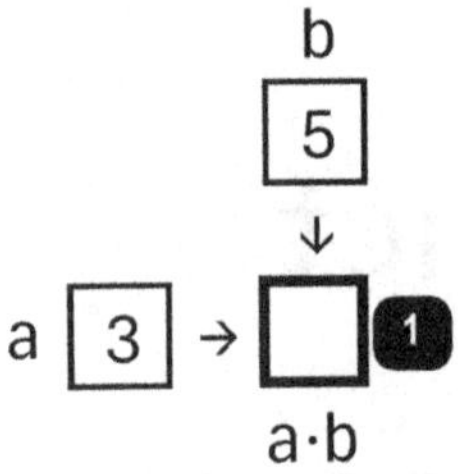

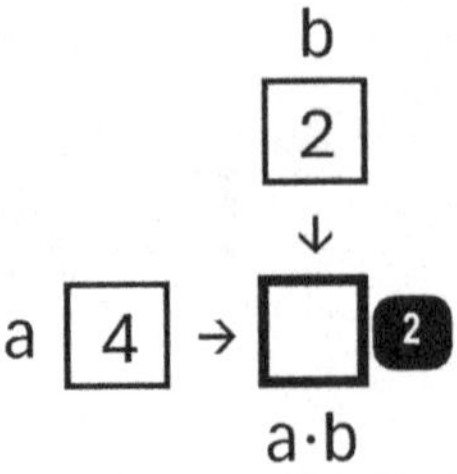

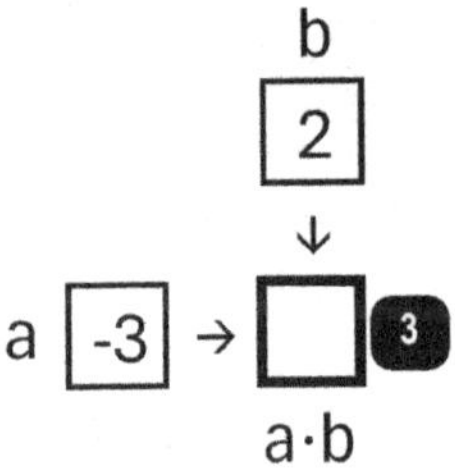

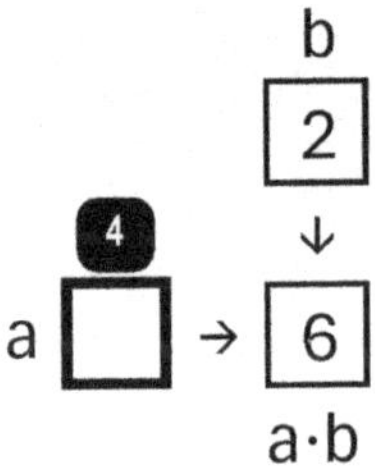

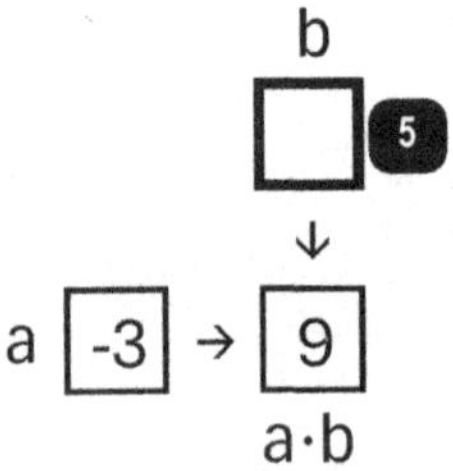

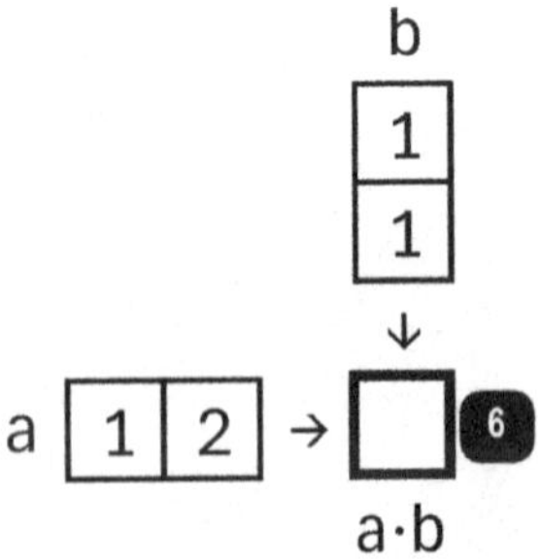

1	2	3	4	5	6
15	8	-6	3	-3	3

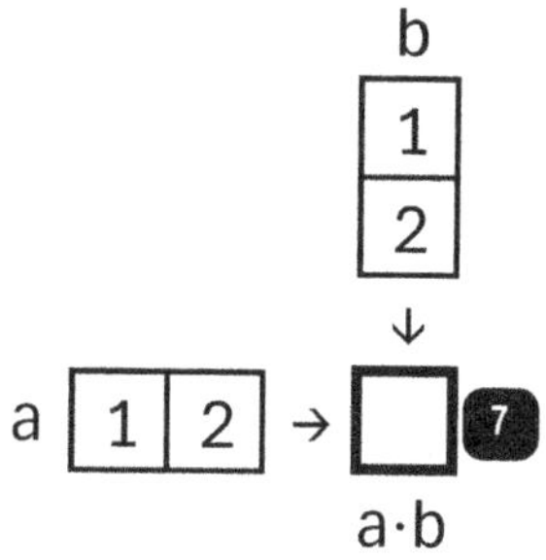

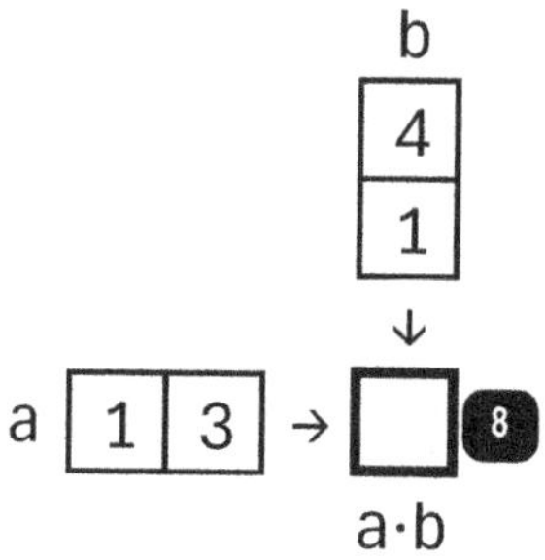

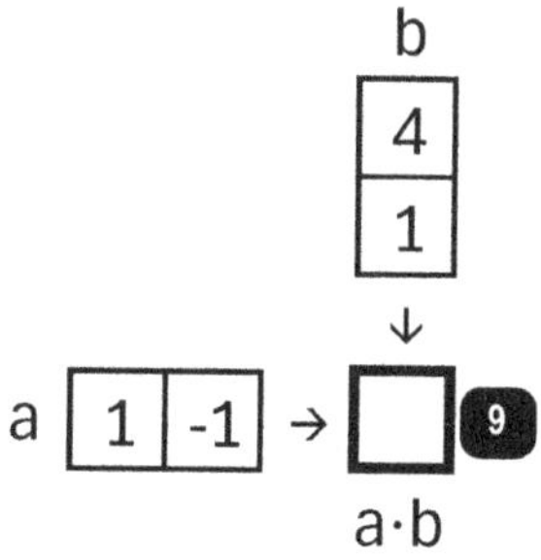

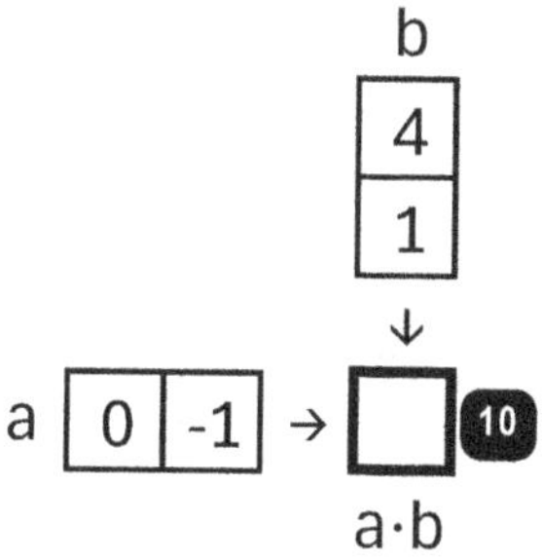

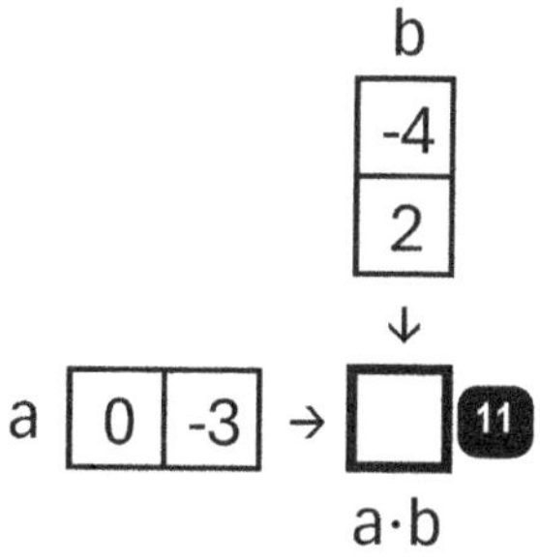

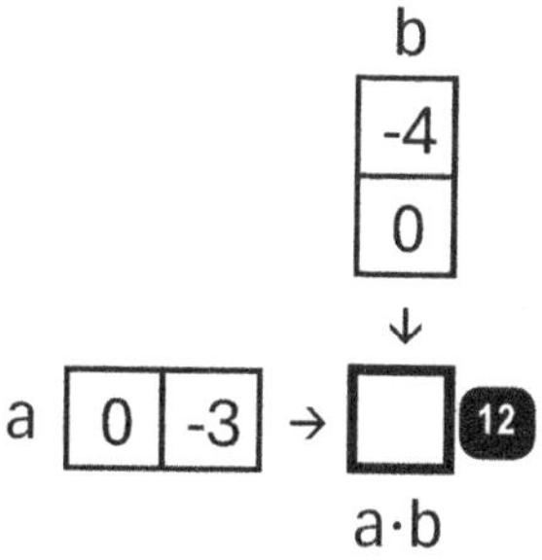

7	8	9	10	11	12
5	7	3	-1	-6	0

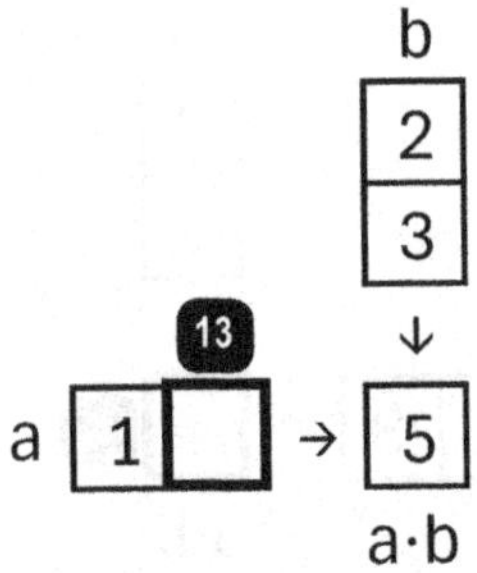

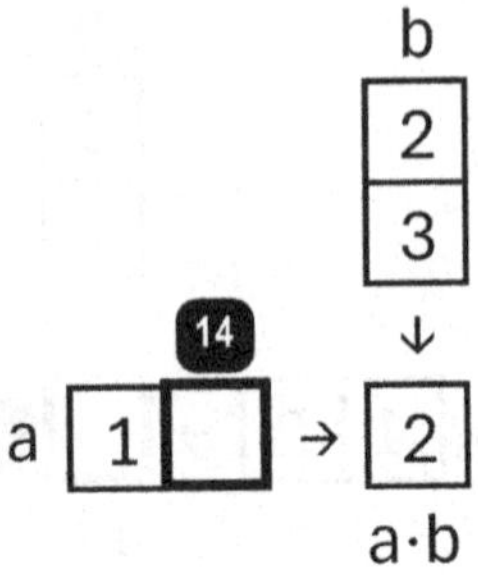

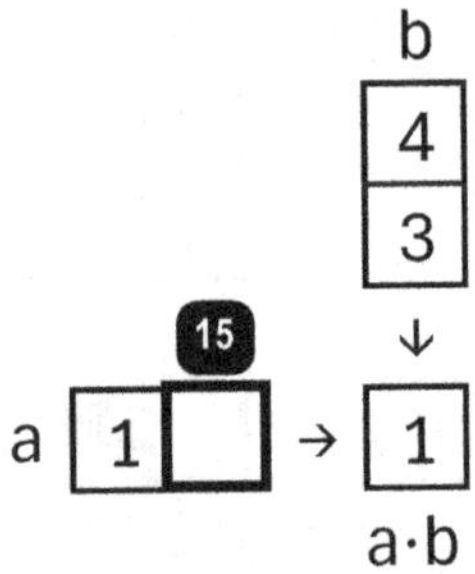

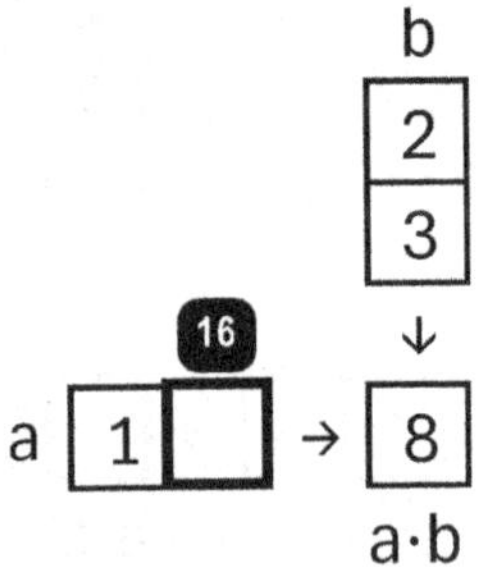

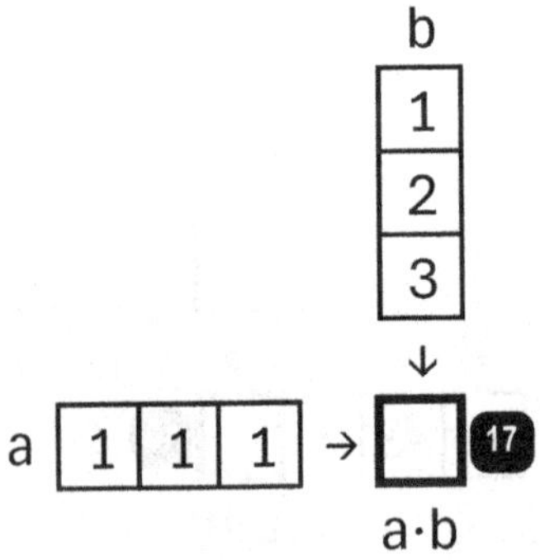

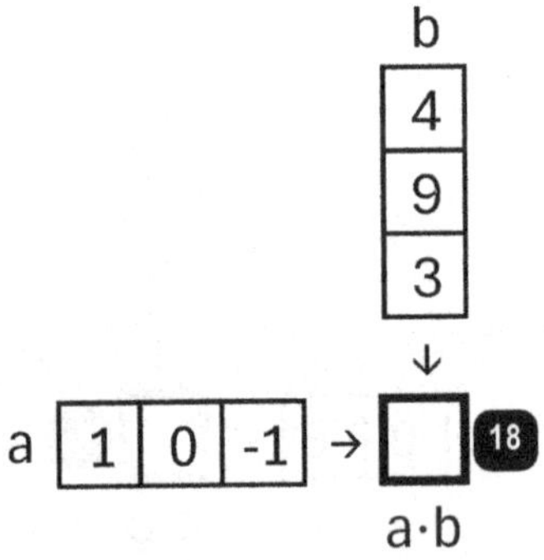

13	14	15	16	17	18
1	0	-1	2	6	1

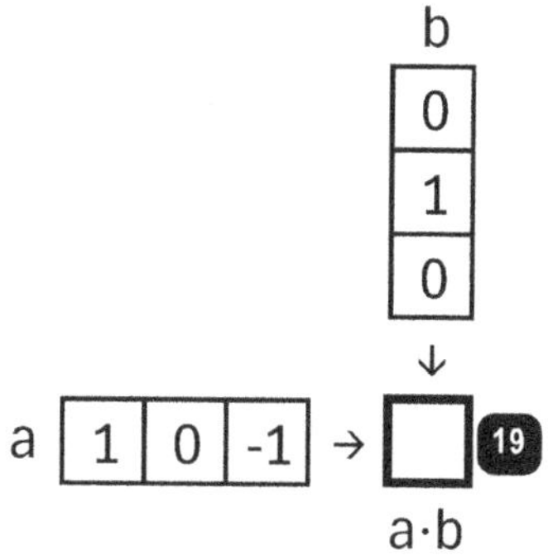

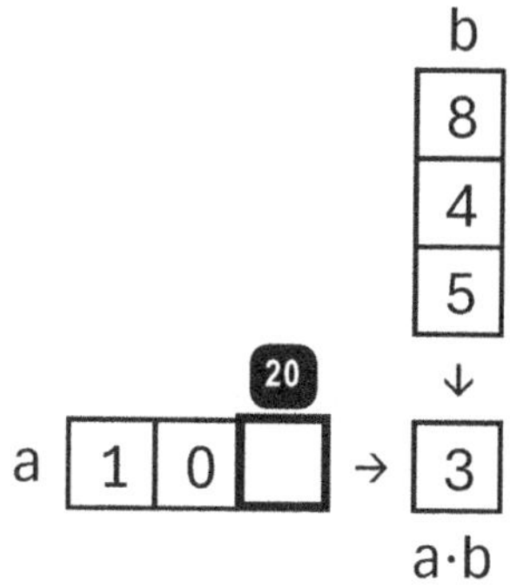

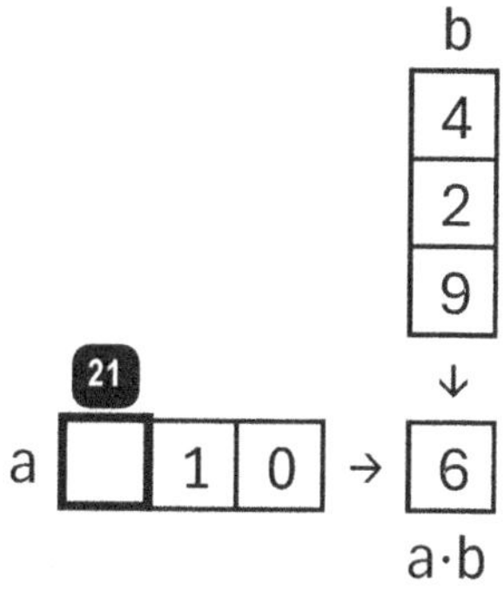

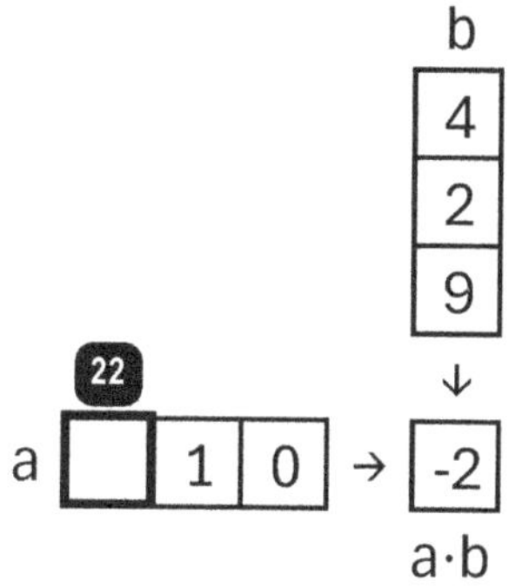

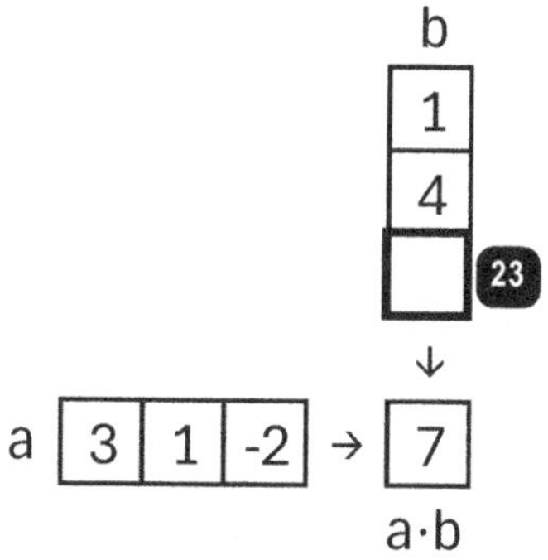

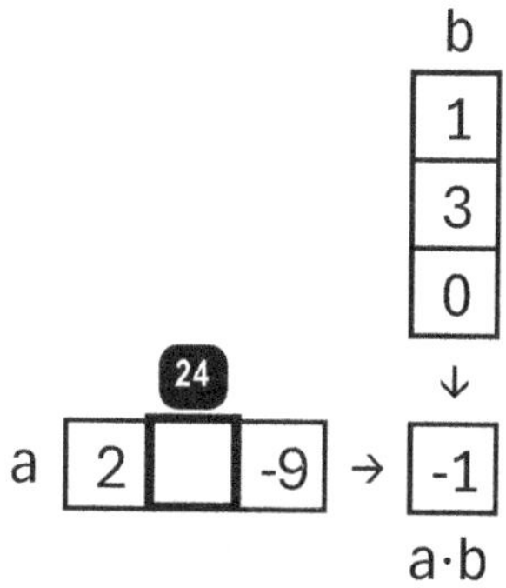

19	20	21	22	23	24
0	-1	1	-1	0	-1

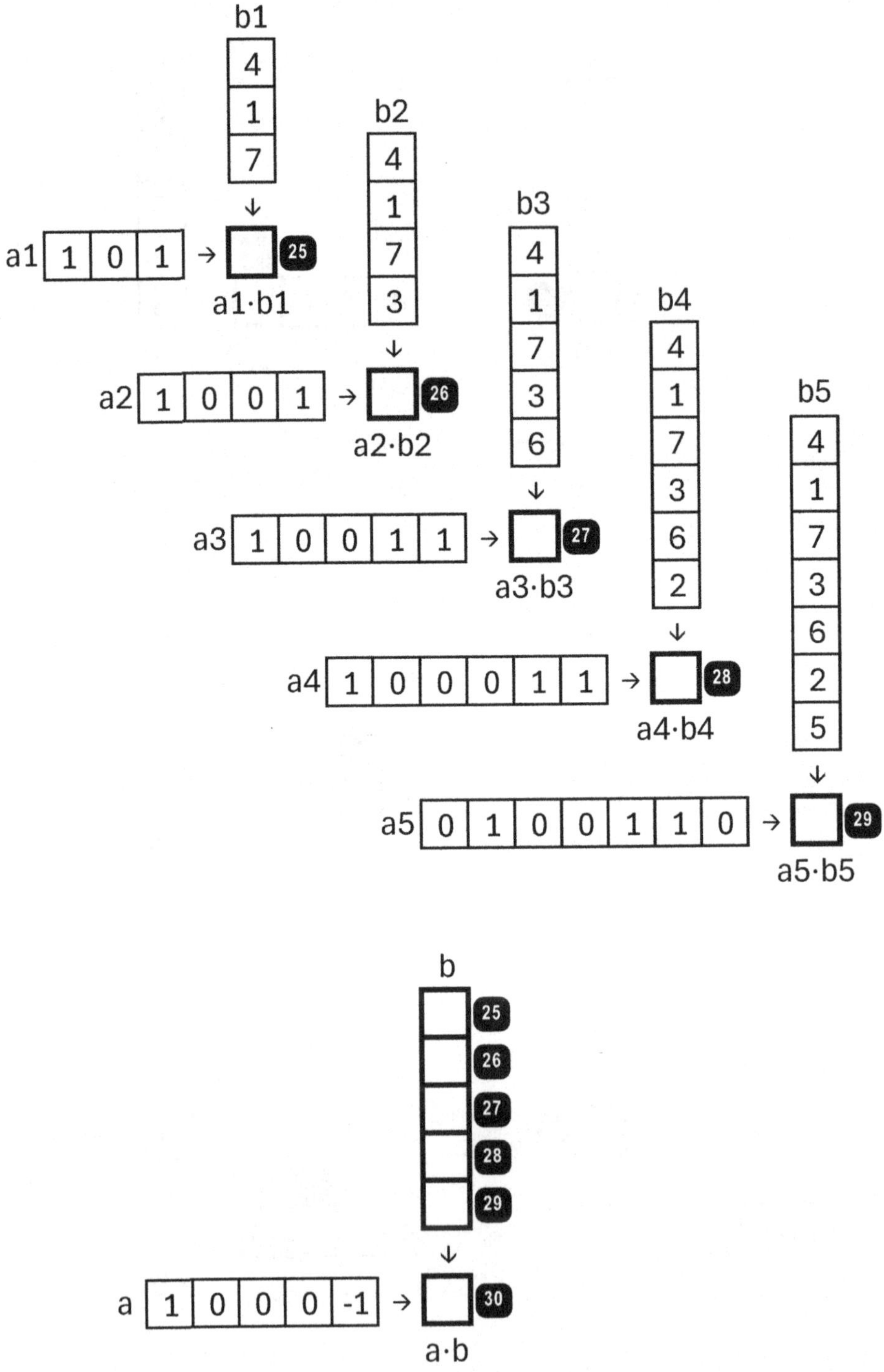

25	26	27	28	29	30
11	7	13	12	9	2

Chapter 2

Matrix Multiplication

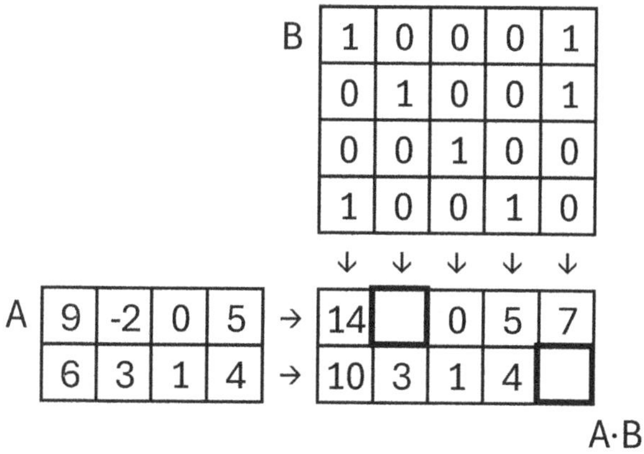

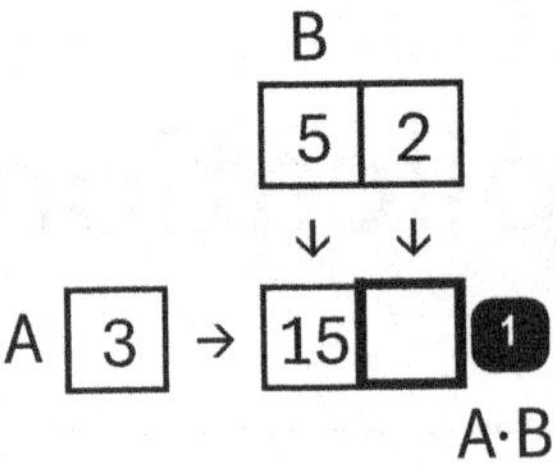

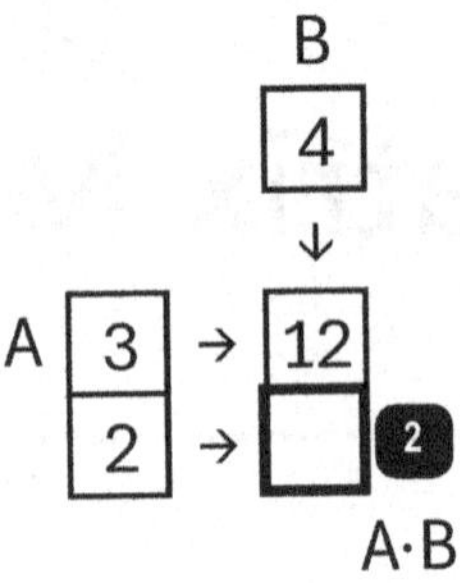

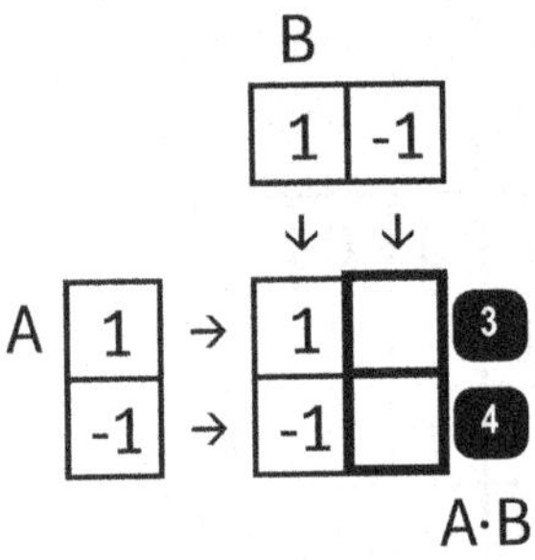

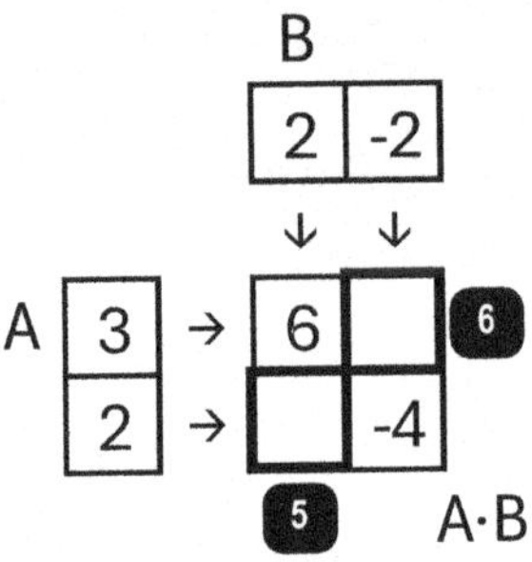

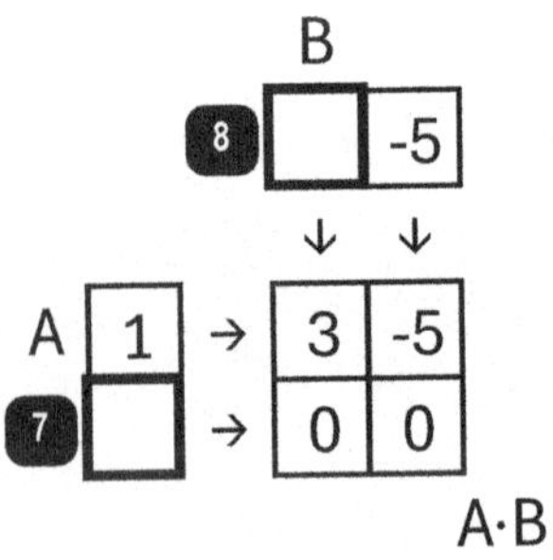

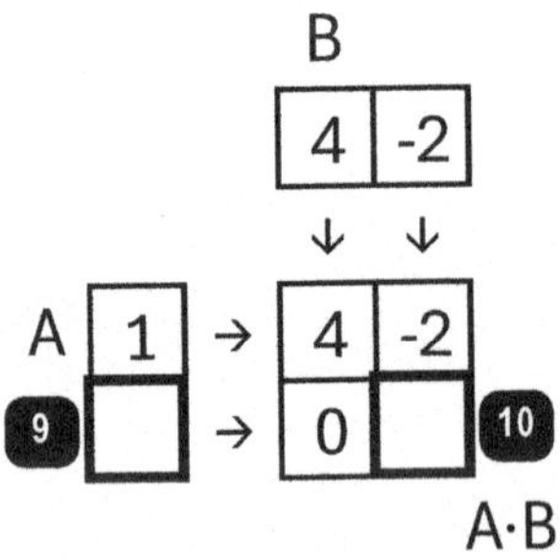

1	2	3	4	5	6	7	8	9	10
6	8	-1	1	4	-6	0	3	0	0

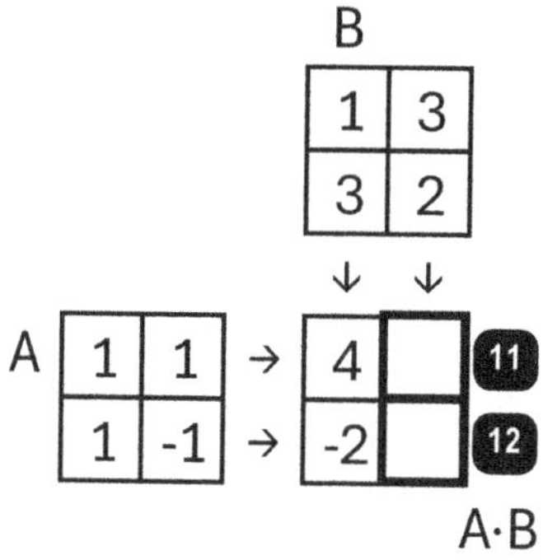

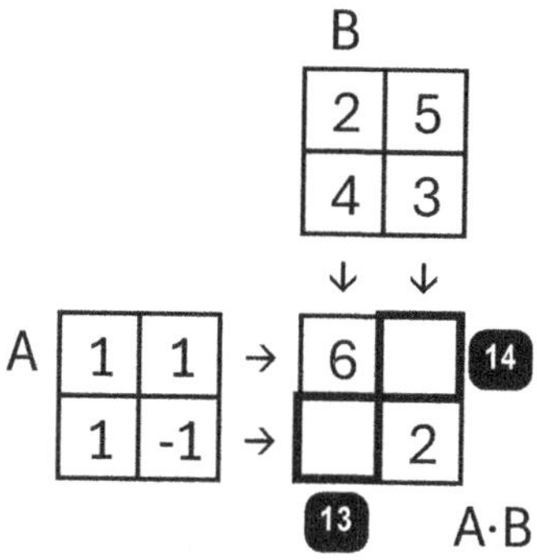

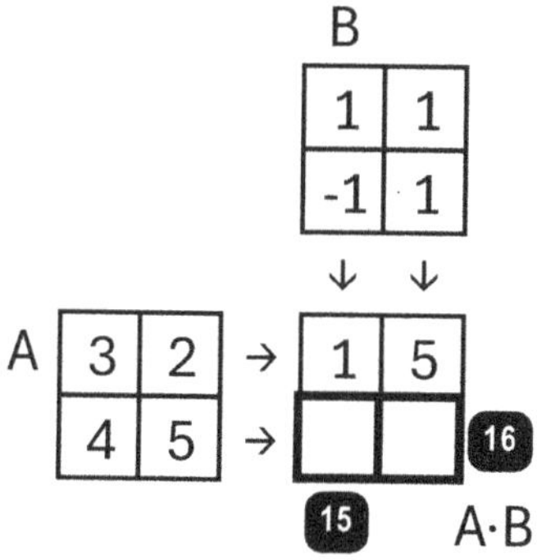

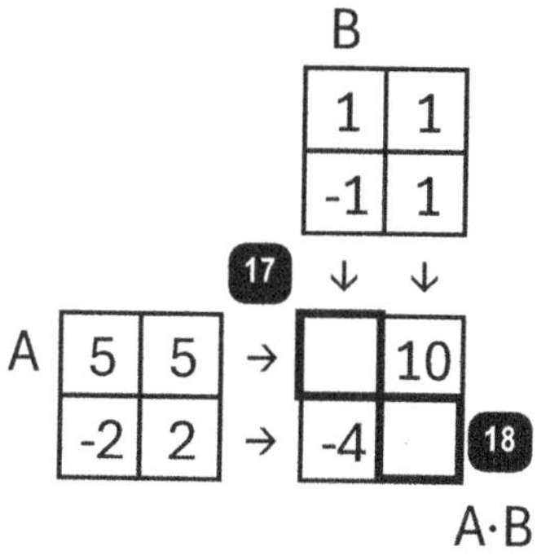

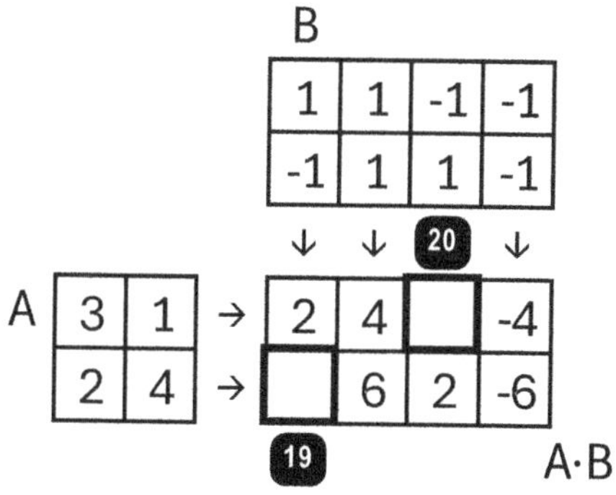

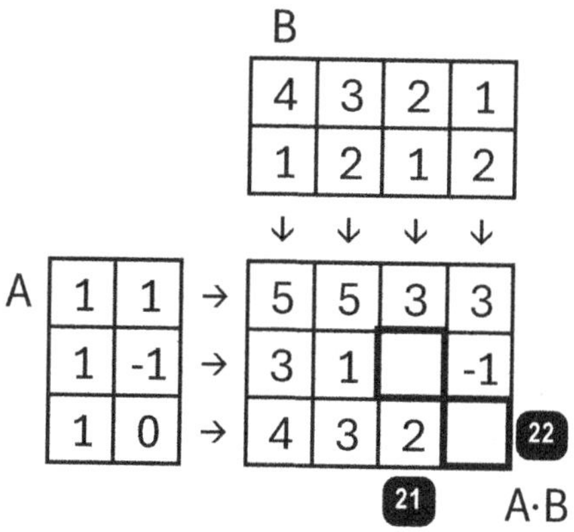

11	12	13	14	15	16	17	18	19	20	21	22
5	1	-2	8	-1	9	0	0	-2	-2	1	1

B

2	1
3	5
4	-1

A

1	0	1
0	1	1

A·B

6	0
23	24

B

1	1
1	0
0	1

A

3	3	-1
2	1	3

A·B

25	26
3	5

A

1	0	1
0	1	1

B

2	1
3	5
4	-1

B·A

2	1	3
3	5	28
27	-1	3

A

3	3	-1
2	1	3

B

1	1
1	-1
1	0

B·A

5	4	30
1	2	-4
3	29	-1

B

3	2
2	1
1	32

A

0	1	31
1	1	1

A·B

2	1
6	1

B

-2	34
4	4
3	1

A

33	0	1
1	1	0

A·B

1	3
2	6

23	24	25	26	27	28	29	30	31	32	33	34
7	4	6	2	4	8	3	2	0	-2	1	2

B

3	2
2	0
1	2
4	3

A

1	1	0	1
1	0	1	-1

A·B

9	36
35	1

A

1	1	0	1
1	0	1	-1

B

3	2
2	0
1	2
4	3

B·A

5	3	2	1
2	37	0	2
3	1	38	-1
7	4	3	1

B

3	2
2	1
1	1
4	3

A

1	1	0	0
0	1	1	0
0	0	1	1
1	0	0	1

A·B

5	3
3	39
40	4
7	5

B

3	2
2	1
1	1
4	3

A

1	-1	0	0
0	1	-1	0
0	0	1	-1
-1	0	0	1

A·B

1	1
1	41
42	-2
1	1

35	36	37	38	39	40	41	42
0	5	2	2	2	5	0	-3

B

1	0	0	0	1
0	1	0	0	1
0	0	1	0	0
1	0	0	1	0

A

9	-2	0	5
6	3	1	4

A·B

14	[43]	0	5	7
10	3	1	4	[44]

B

0	0	0	1	1
1	0	1	0	1
0	1	1	0	1
0	0	0	1	0

A

1	1	1	0
0	1	1	1
1	0	1	0
1	1	0	1
1	1	-1	0
[49]	1	0	1
1	0	[50]	0

A·B

1	[45]	2	1	3
1	1	2	1	[46]
0	1	1	1	2
1	0	[47]	2	2
1	-1	0	[48]	1
1	0	1	1	1
0	1	1	1	2

43	44	45	46	47	48	49	50
-2	9	1	2	1	1	0	1

Chapter 3

Linear Layer

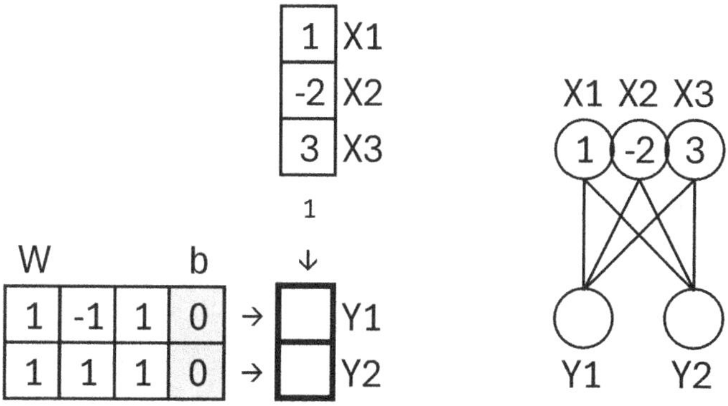

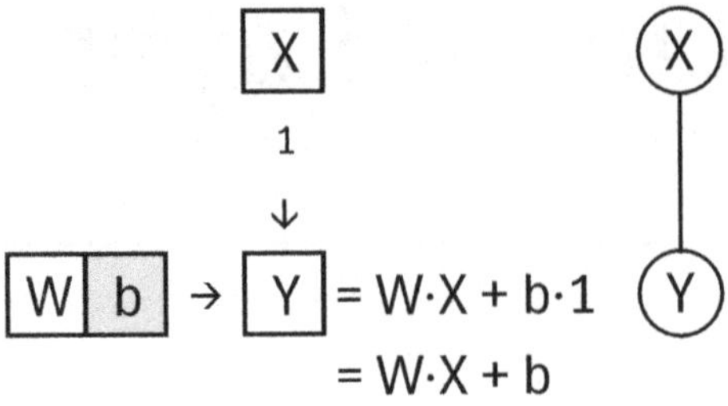

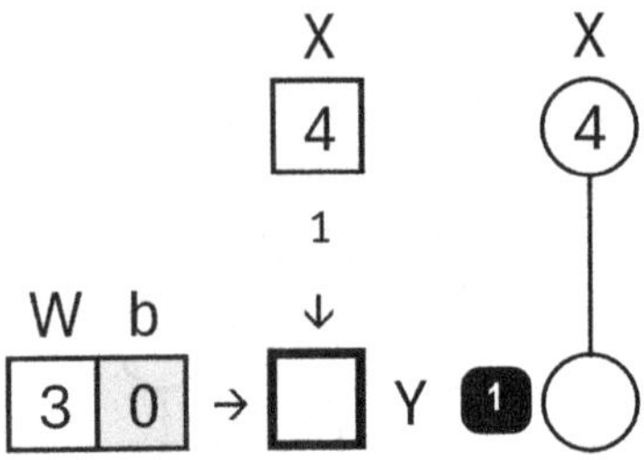

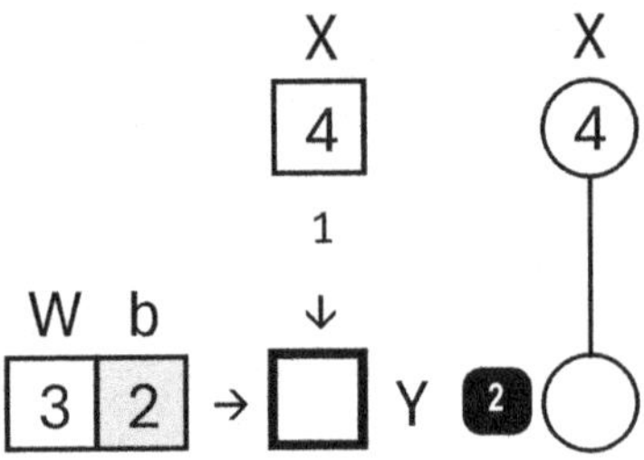

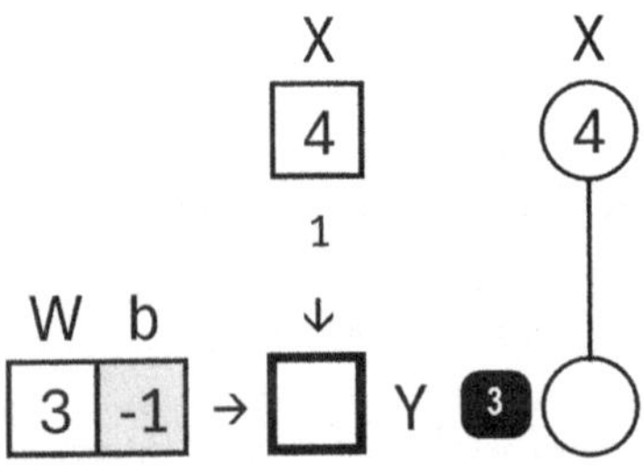

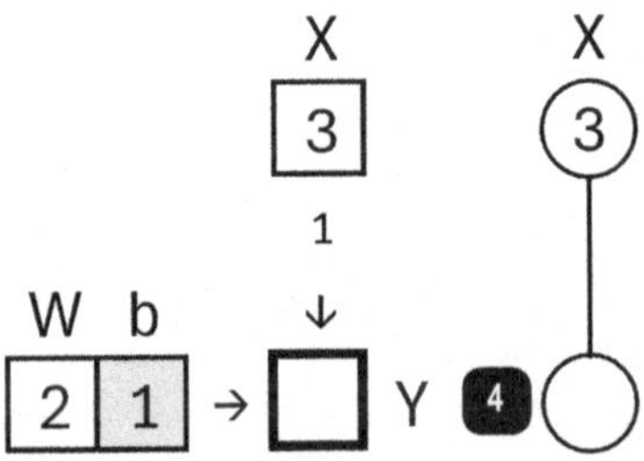

1	2	3	4
12	14	11	7

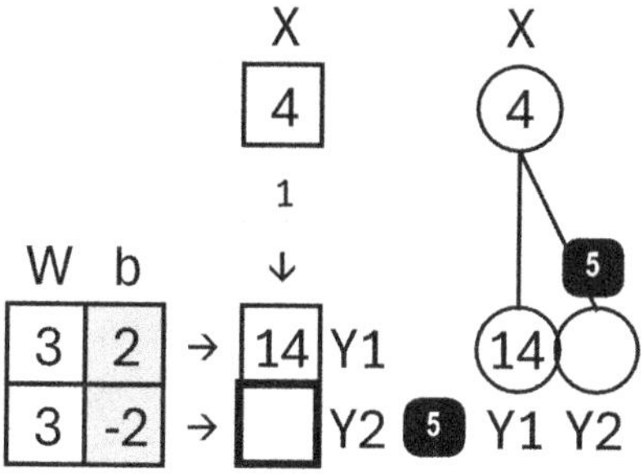

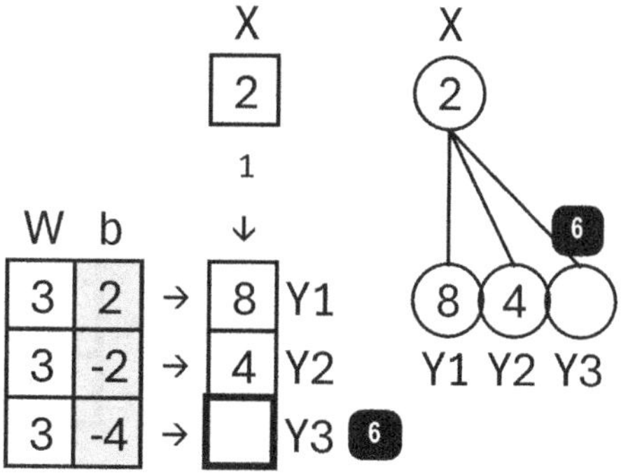

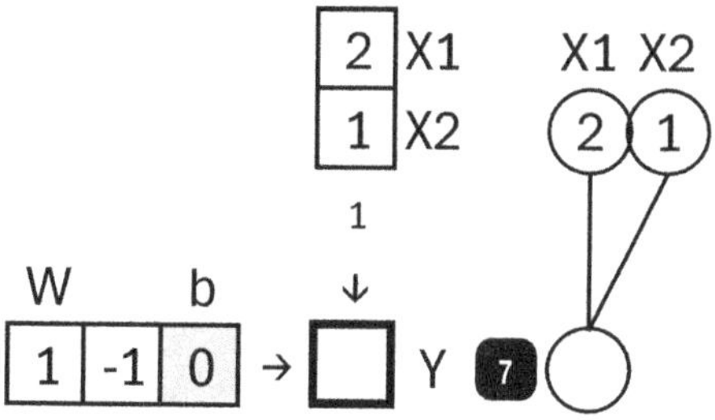

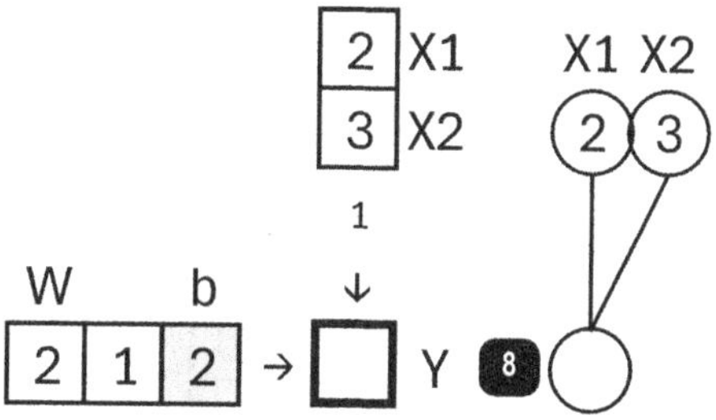

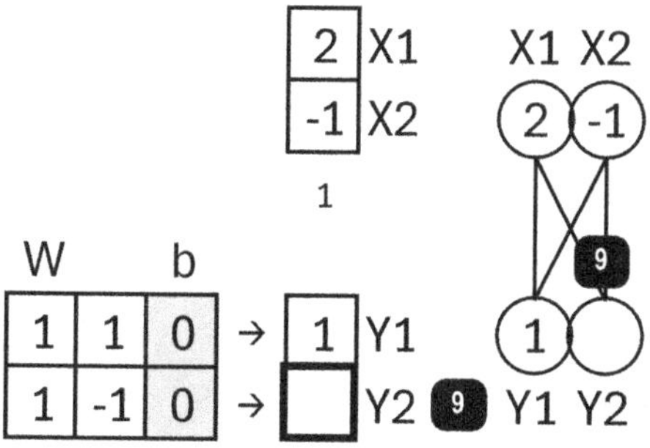

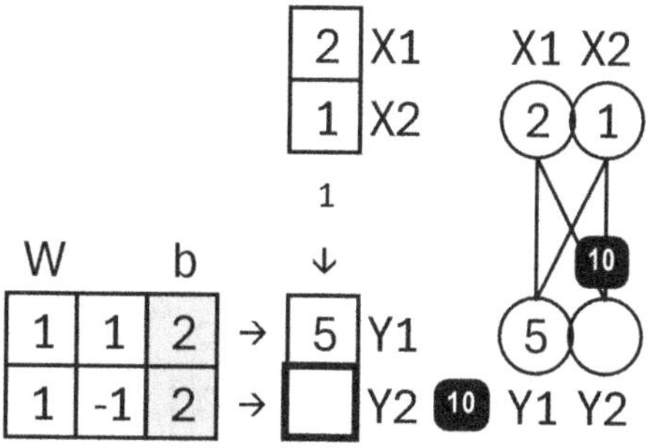

5	6	7	8	9	10
10	2	1	9	3	3

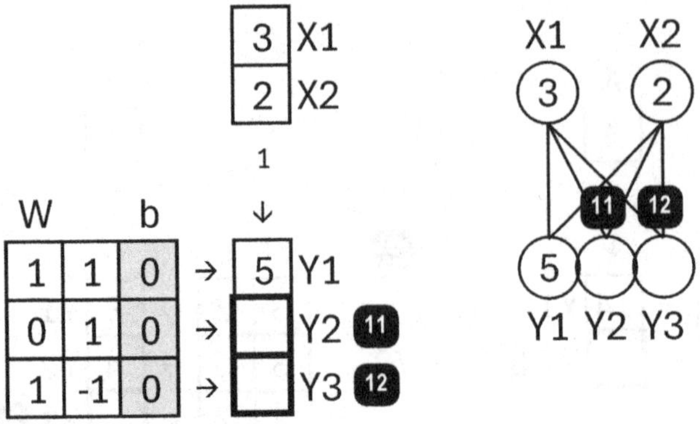

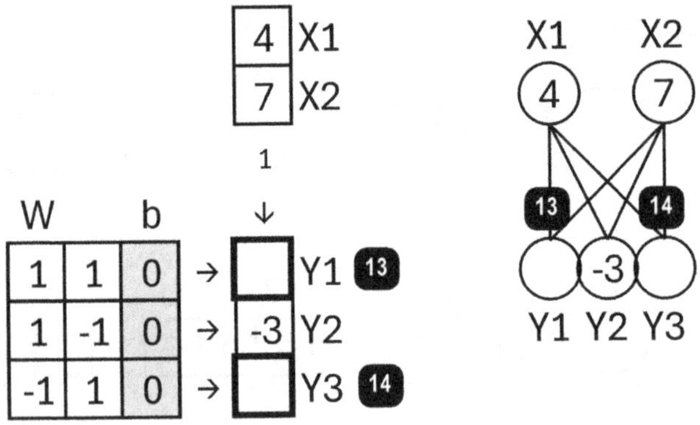

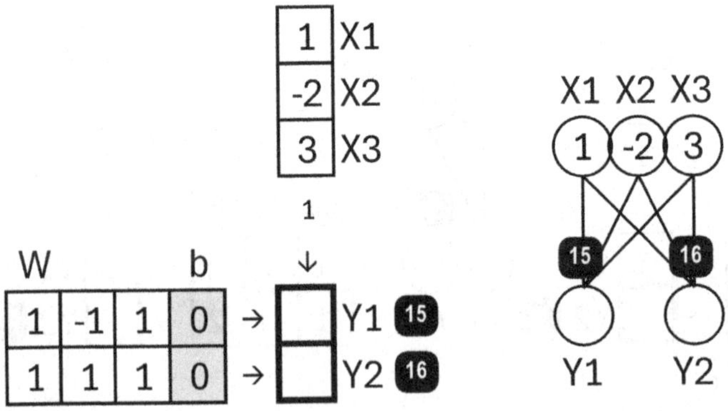

11	12	13	14	15	16
2	1	11	3	6	2

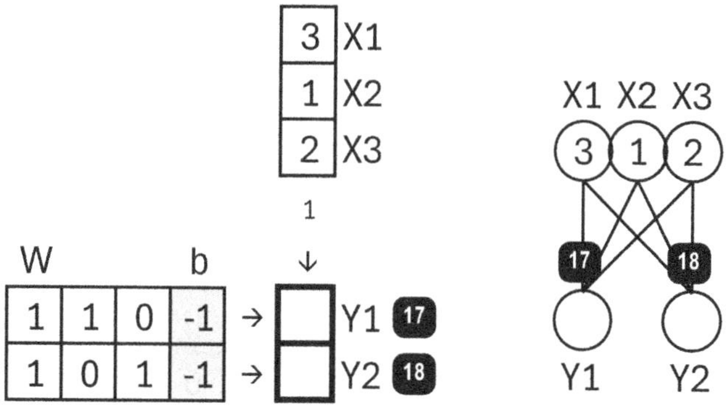

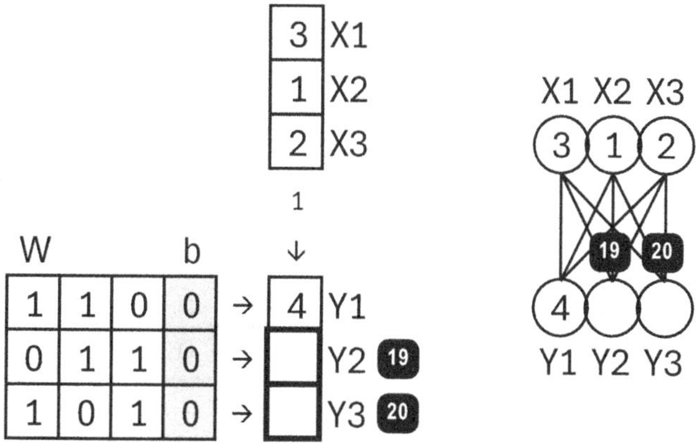

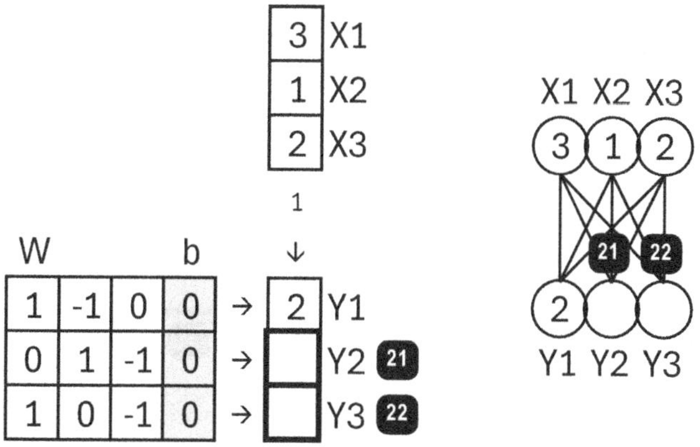

17	18	19	20	21	22
3	4	3	5	-1	1

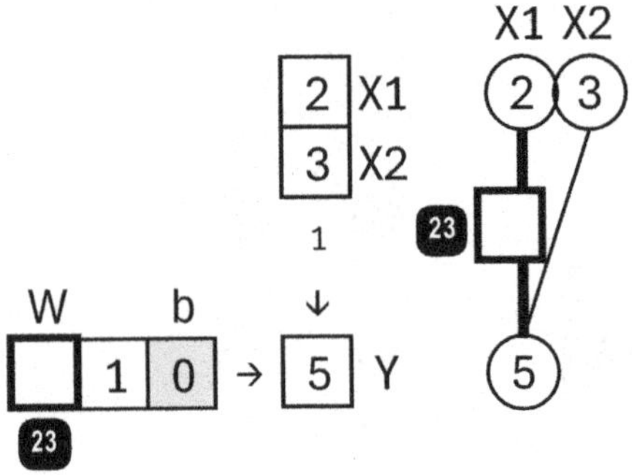

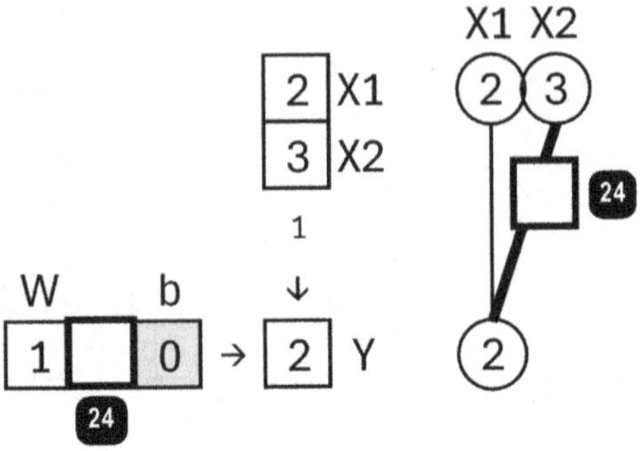

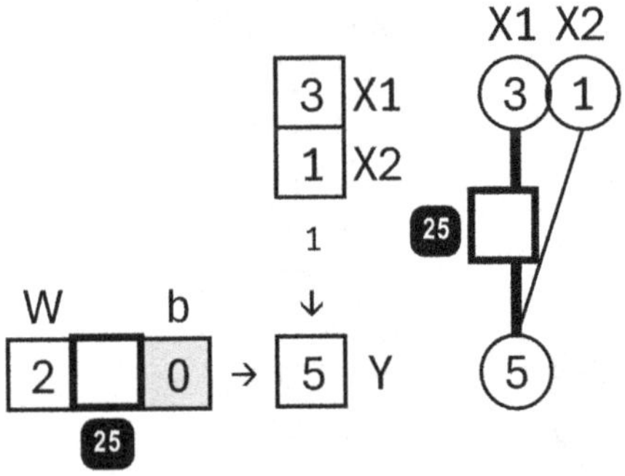

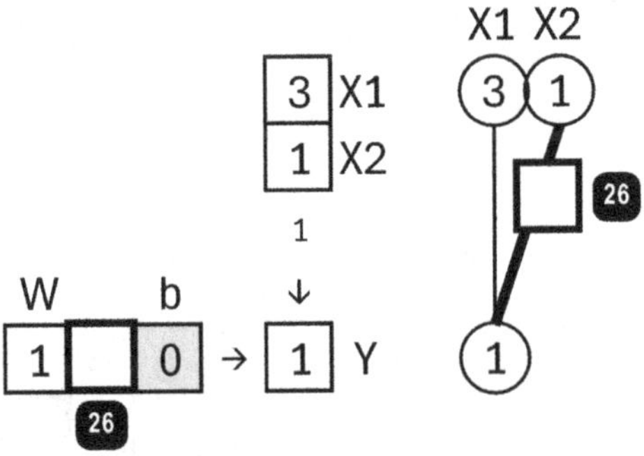

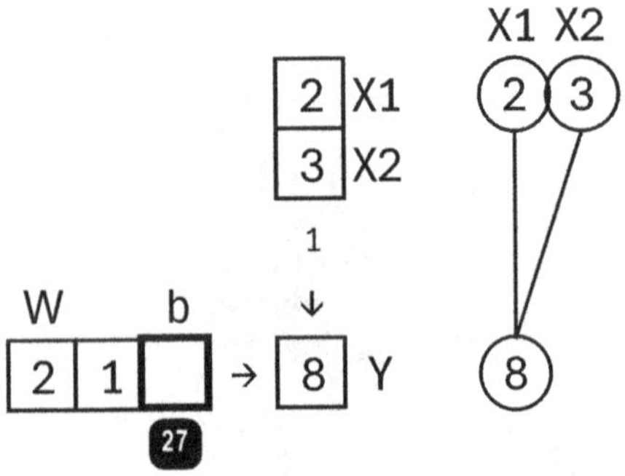

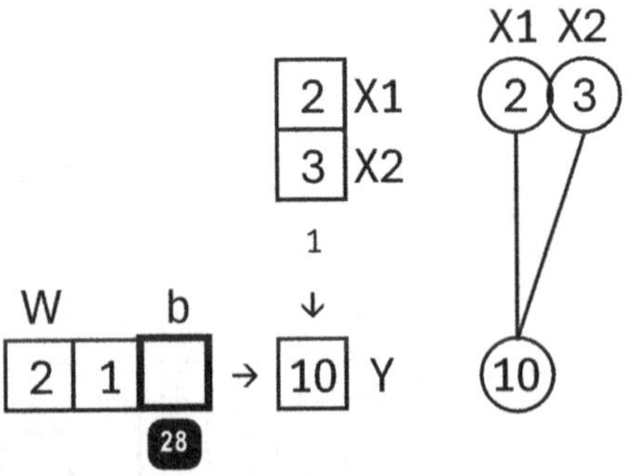

23	24	25	26	27	28
1	0	-1	-2	1	3

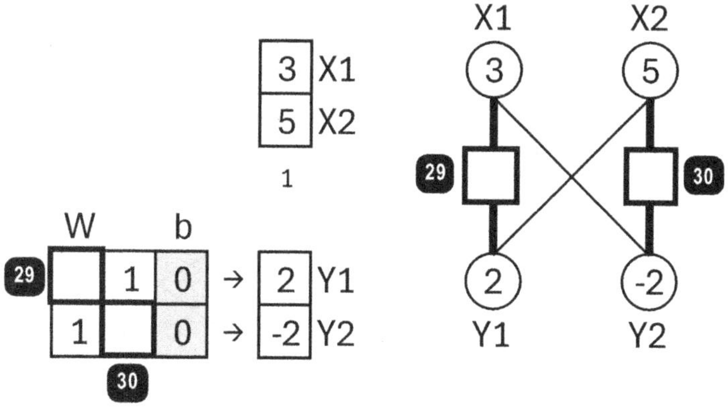

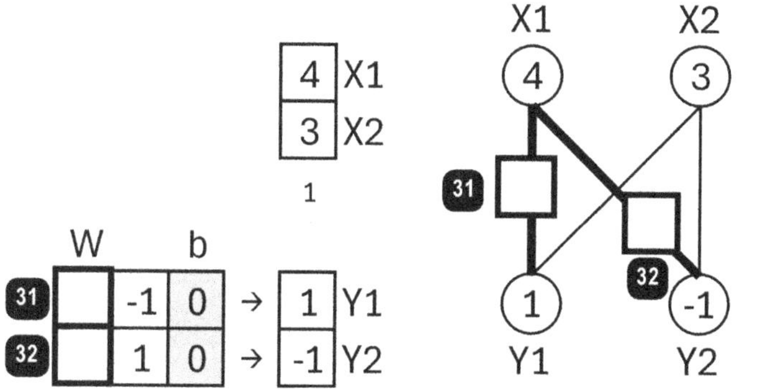

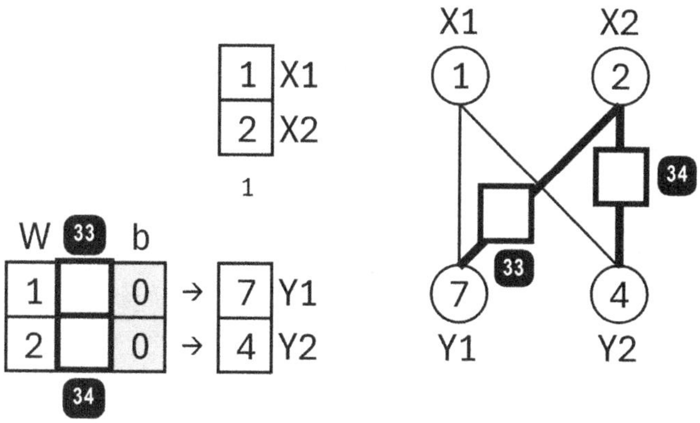

29	30	31	32	33	34
-1	-1	1	-1	3	1

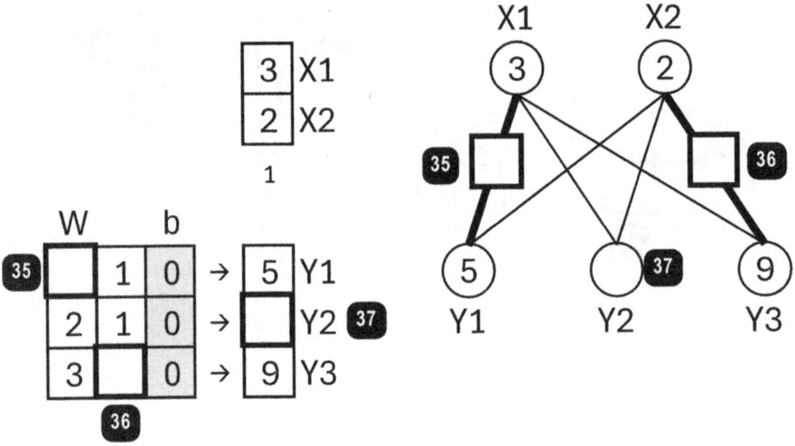

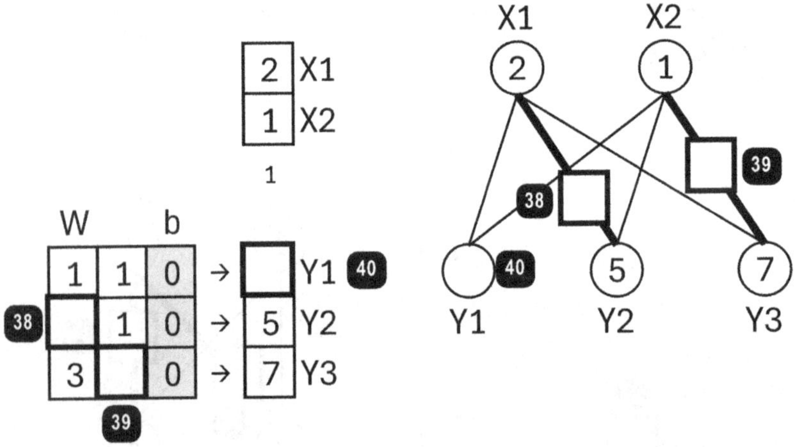

35	36	37	38	39	40
1	0	8	2	1	3

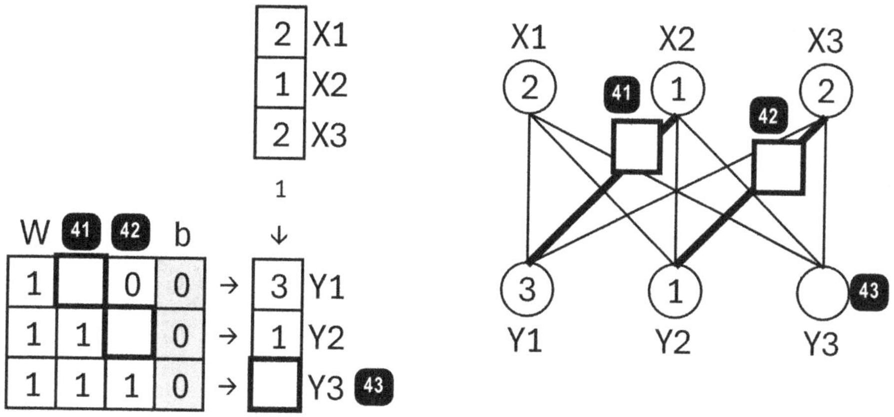

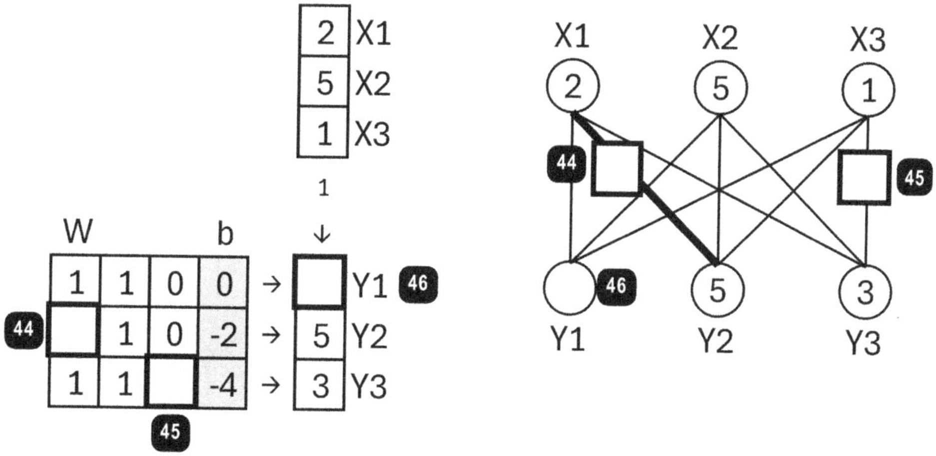

41	42	43	44	45	46
1	-1	5	1	0	7

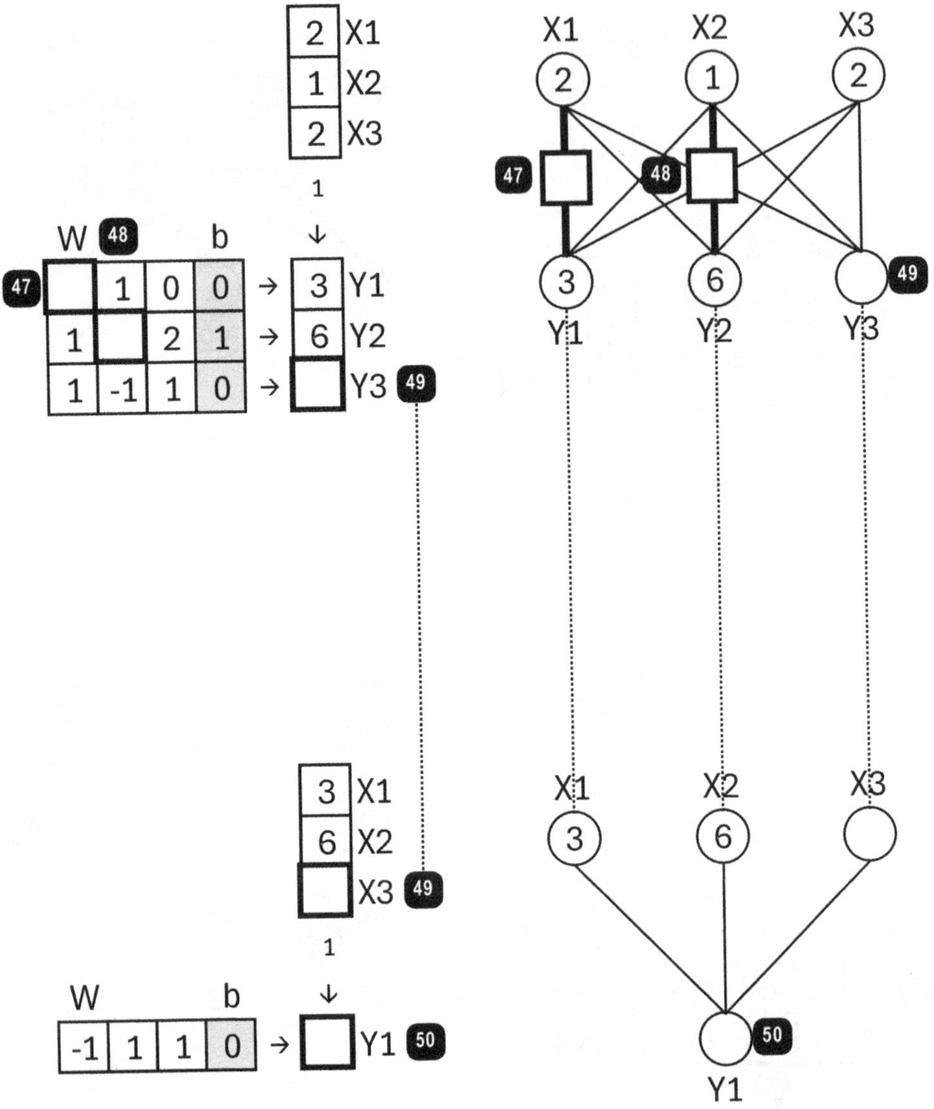

47	48	49	50
1	-1	3	6

Chapter 4

Activation

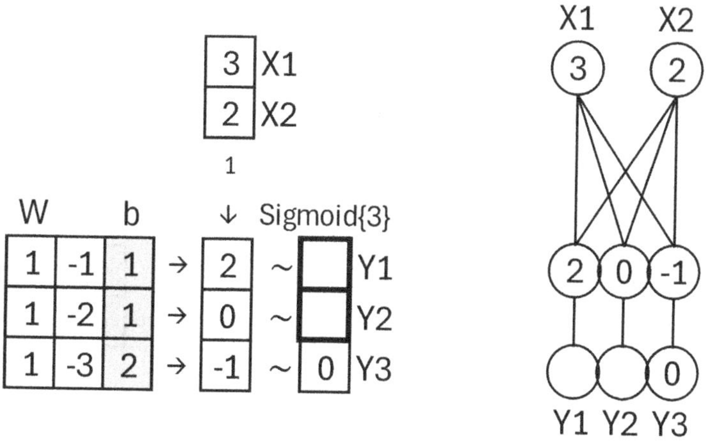

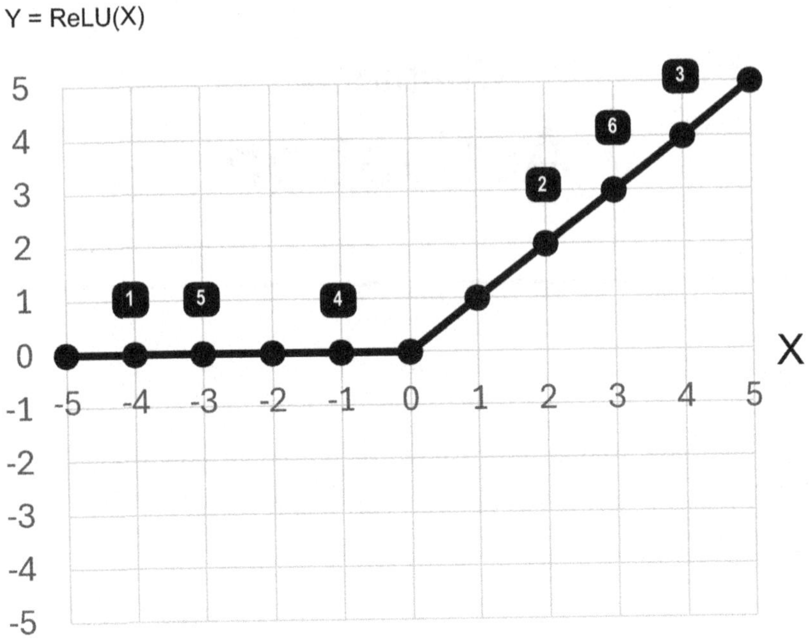

ReLU

X		Y	
-5	~	0	
-4	~		1
-3	~	0	
-2	~	0	
-1	~	0	
0	~	0	
1	~	1	
2	~		2
3	~	3	
4	~		3
5	~	5	

ReLU

X		Y	
4	~	4	
-3	~	0	
-3	~	0	
1	~	1	
5	~	5	
0	~	0	
3	~	3	
-1	~		4
-1	~	0	
-3	~		5
3	~		6

1	2	3	4	5	6
0	2	4	0	0	3

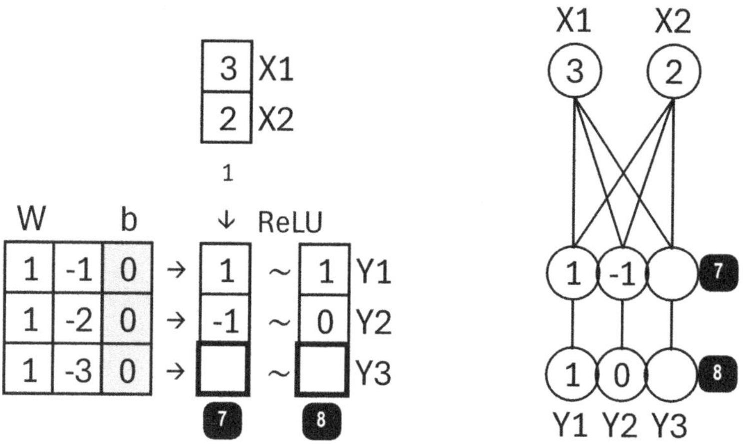

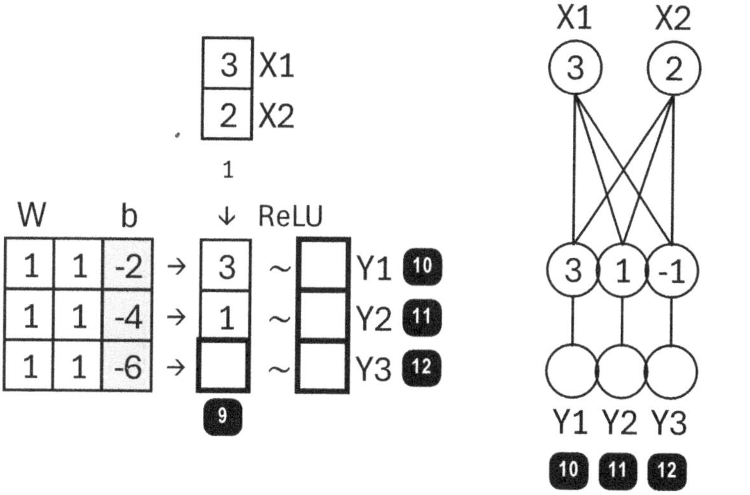

7	8	9	10	11	12
-3	0	-1	3	1	0

Y = Sigmoid{3}   3 Level Quantization

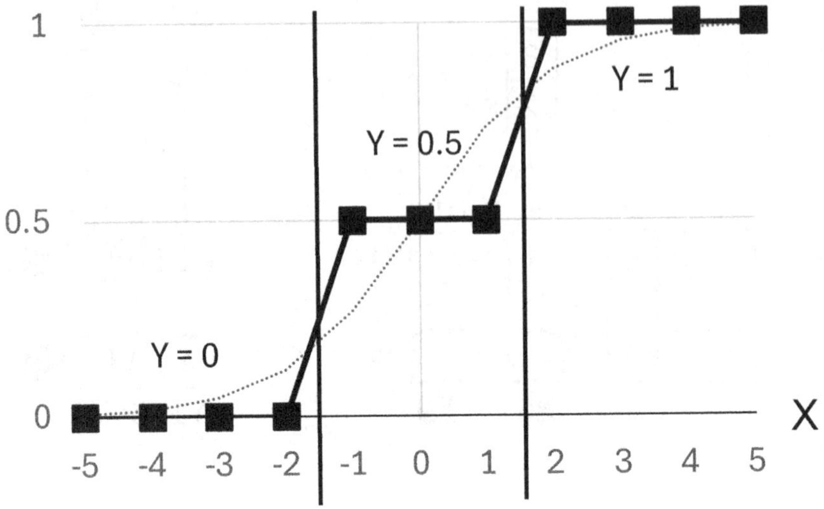

Sigmoid{3}

X		Y	
-5	~	0	
-4	~		13
-3	~		14
-2	~	0	
-1	~	0.5	
0	~		15
1	~	0.5	
2	~	1	
3	~		16
4	~	1	
5	~	1	

Sigmoid{3}

X		Y	
0	~	0.5	
-3	~		17
-2	~	0	
1	~		18
5	~	1	
0	~	0.5	
2	~	1	
-1	~	0.5	
-1	~		19
-3	~	0	
3	~		20

13	14	15	16	17	18	19	20
0	0	0.5	1	0	0.5	0.5	1

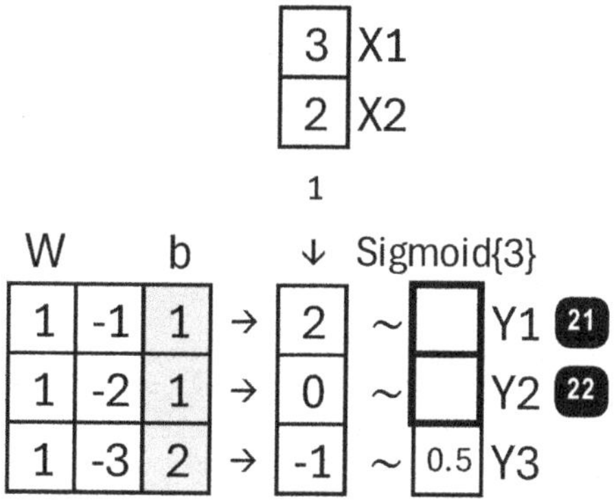

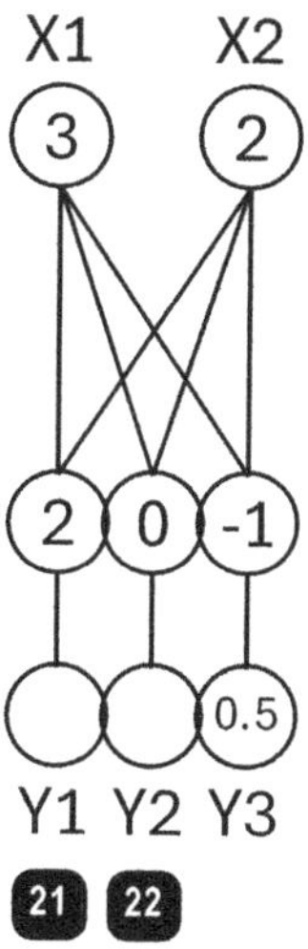

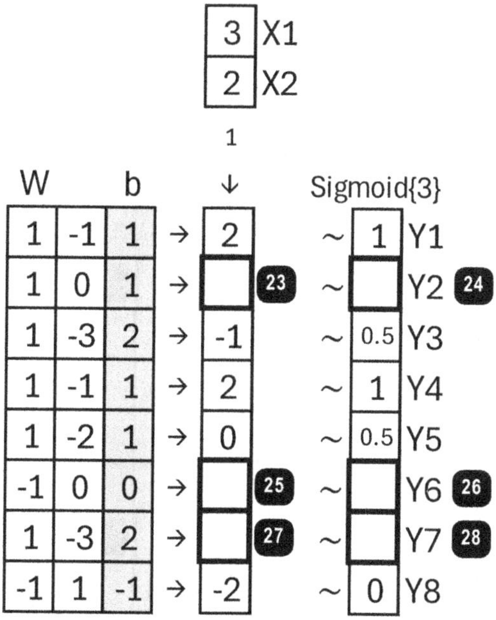

21	22	23	24	25	26	27	28
1	0.5	4	1	-3	0	-1	0.5

Y = Tanh{3}(X) 3 Level Quantization

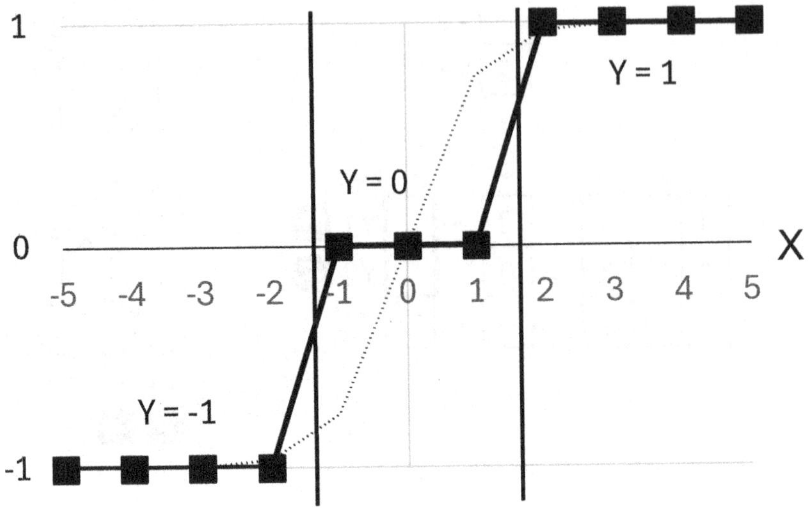

Tanh{3}

X		Y	
-5	~	-1	
-4	~		29
-3	~		30
-2	~	-1	
-1	~	0	
0	~		31
1	~	0	
2	~	1	
3	~		32
4	~	1	
5	~	1	

Tanh{3}

X		Y	
0	~	0	
-3	~		33
-2	~	-1	
1	~		34
5	~	1	
0	~	0	
2	~	1	
-1	~	0	
-1	~		35
-3	~	-1	
3	~		36

29	30	31	32	33	34	35	36
-1	-1	0	1	-1	0	0	1

Y = Sigmoid{5}(X)   5 Level Quantization

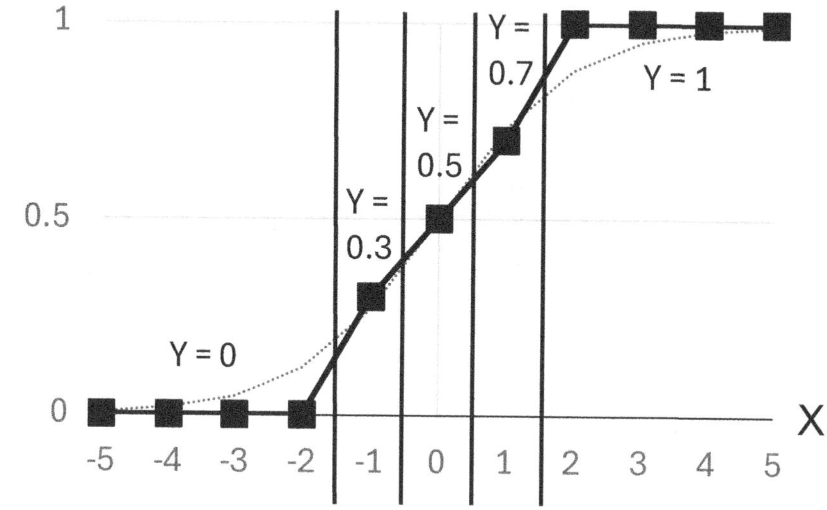

Sigmoid{5}

X		Y	
-5	~	0	
-4	~		37
-3	~		38
-2	~	0	
-1	~	0.3	
0	~		39
1	~	0.7	
2	~	1	
3	~		40
4	~	1	
5	~	1	

Sigmoid{5}

X		Y	
0	~	0.5	
-3	~		41
-2	~	0	
1	~		42
5	~	1	
0	~	0.5	
2	~	1	
-1	~	0.3	
-1	~		43
-3	~	0	
3	~		44

37	38	39	40	41	42	43	44
0	0	0.5	1	0	0.7	0.3	1

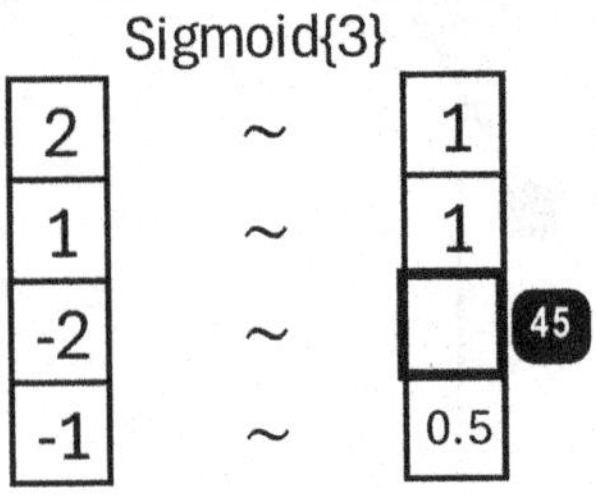

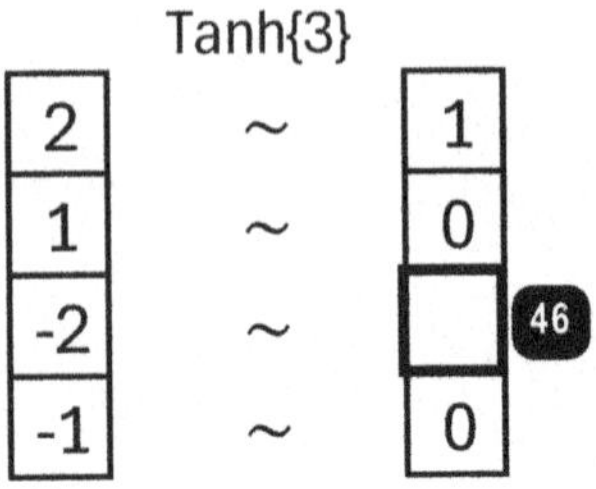

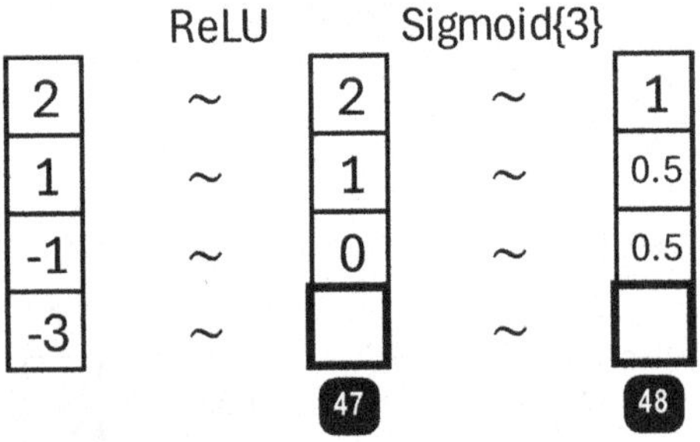

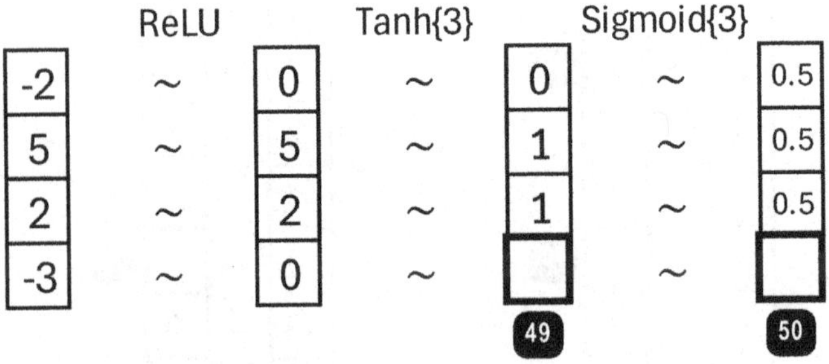

45	46	47	48	49	50
0	-1	0	0.5	0	0.5

Chapter 5

Artificial Neuron

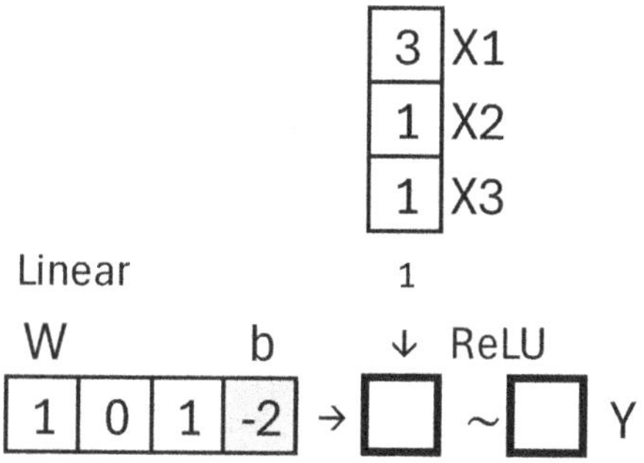

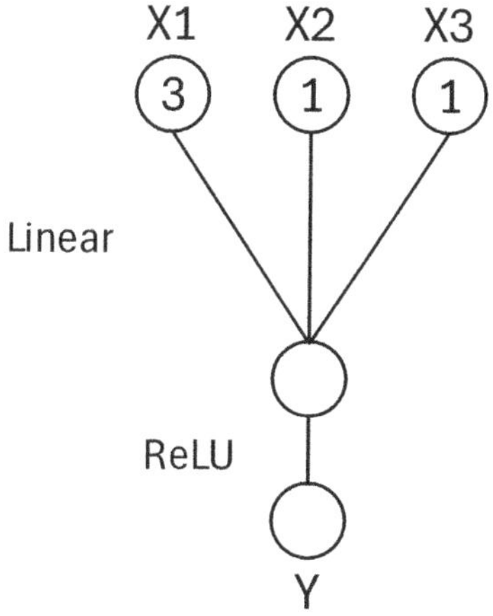

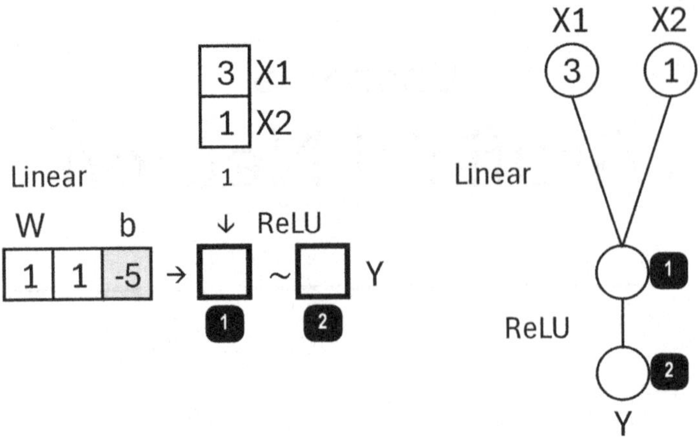

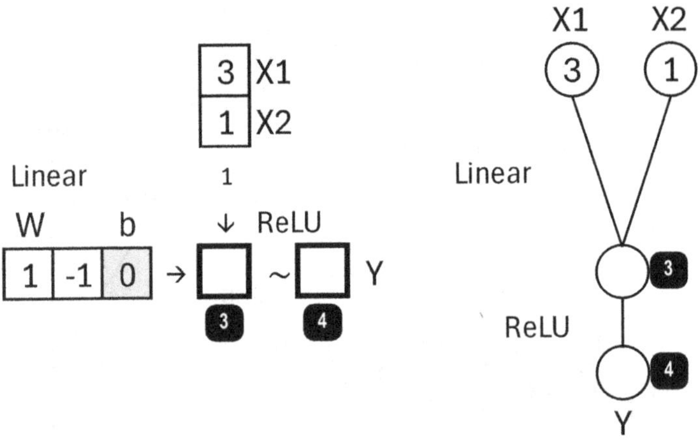

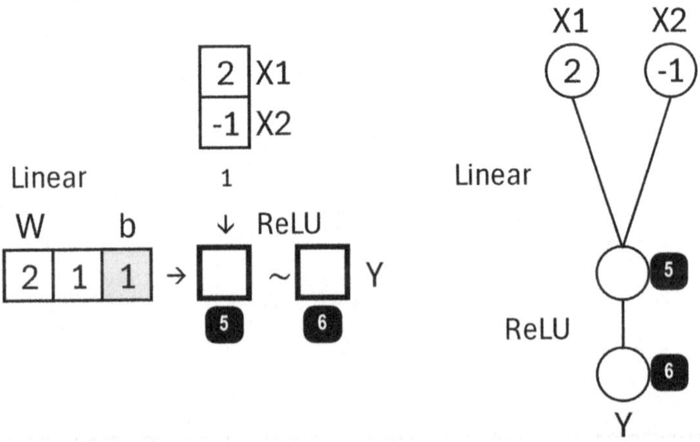

1	2	3	4	5	6
-1	0	2	2	4	4

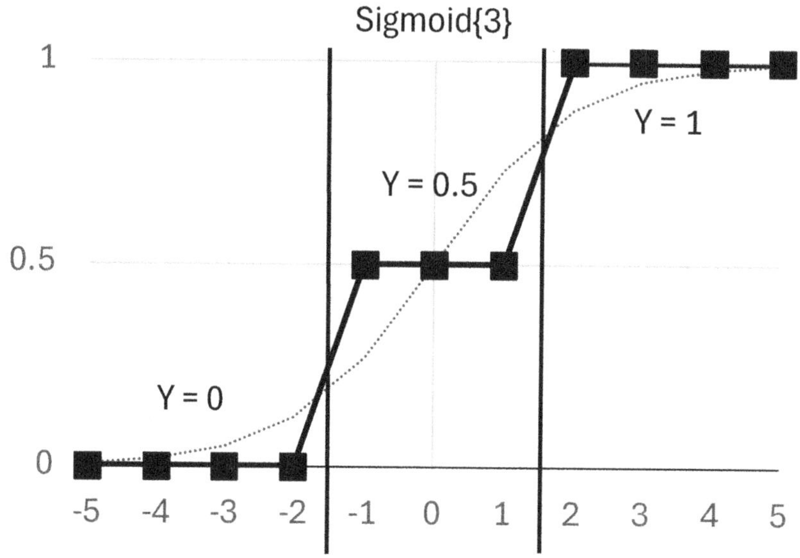
Sigmoid{3}
1
0.5
0
-5
-4
-3
-2
-1
0
1
2
3
4
5
Y = 0
Y = 0.5
Y = 1

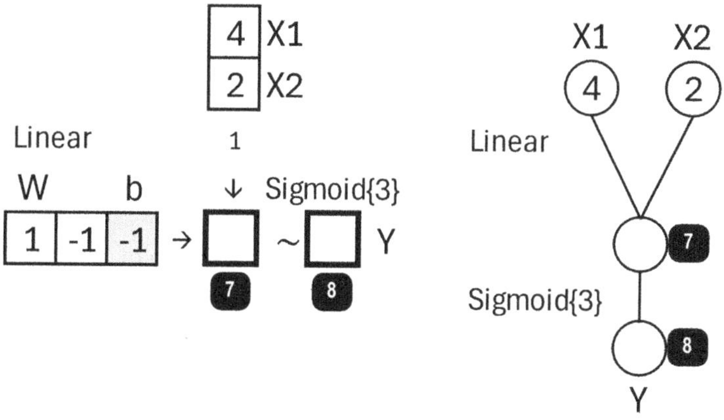
4
X1
2
X2
Linear
1
W
b
Sigmoid{3}
1
-1
-1
Y
7
8
X1
X2
4
2
Linear
7
Sigmoid{3}
8
Y

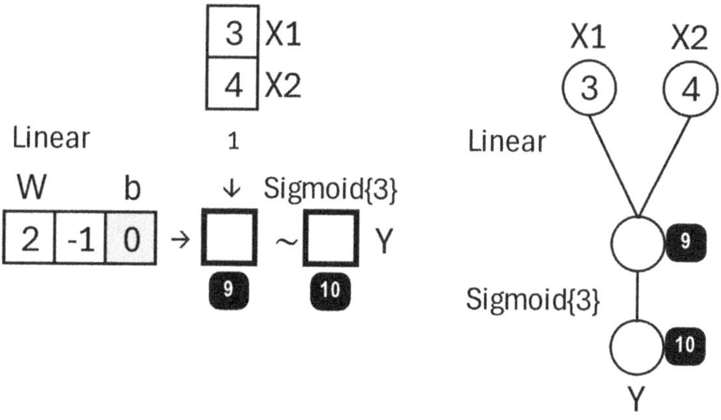
3
X1
4
X2
Linear
1
W
b
Sigmoid{3}
2
-1
0
Y
9
10
X1
X2
3
4
Linear
9
Sigmoid{3}
10
Y

7	8	9	10
1	0.5	2	1

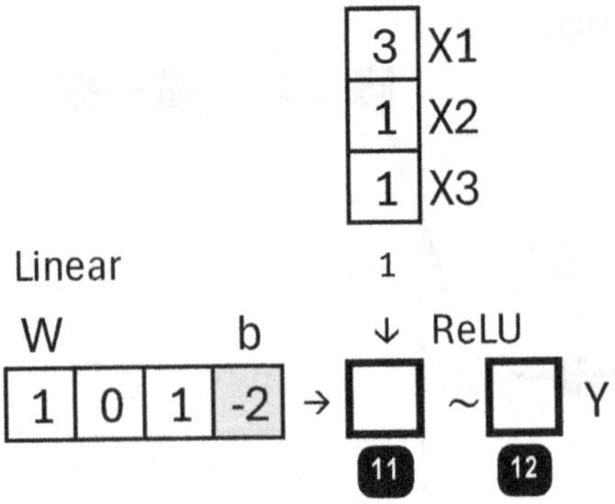

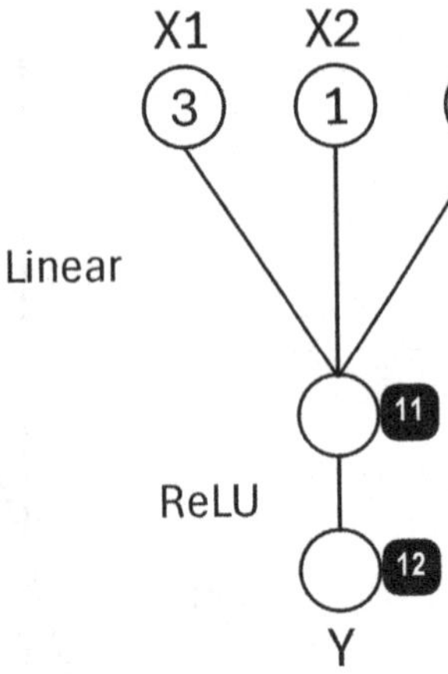

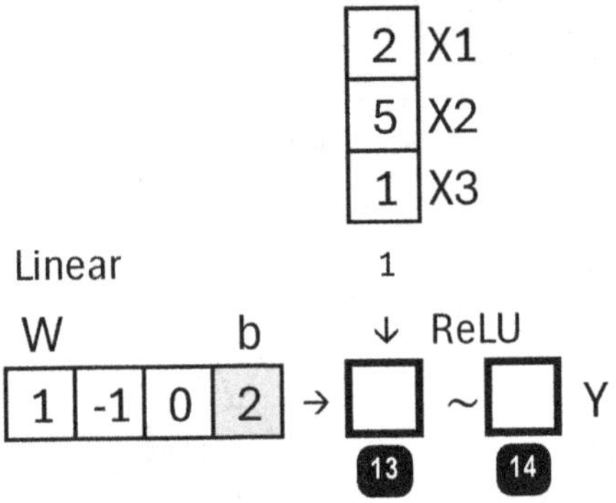

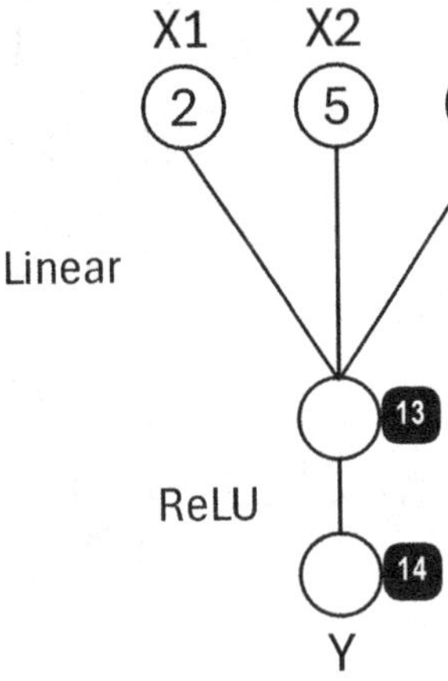

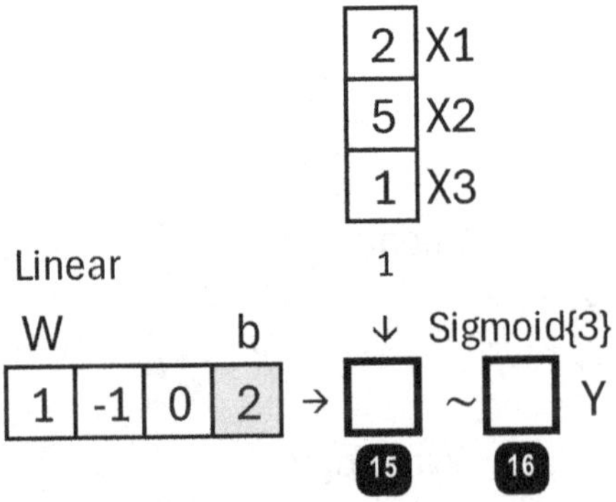

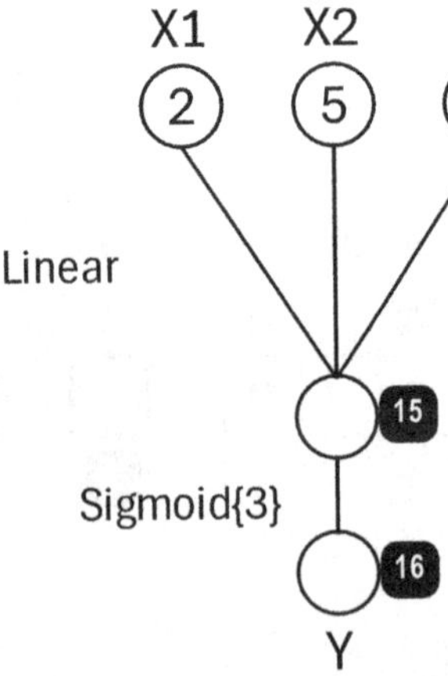

11	12	13	14	15	16
2	2	-1	0	-1	0.5

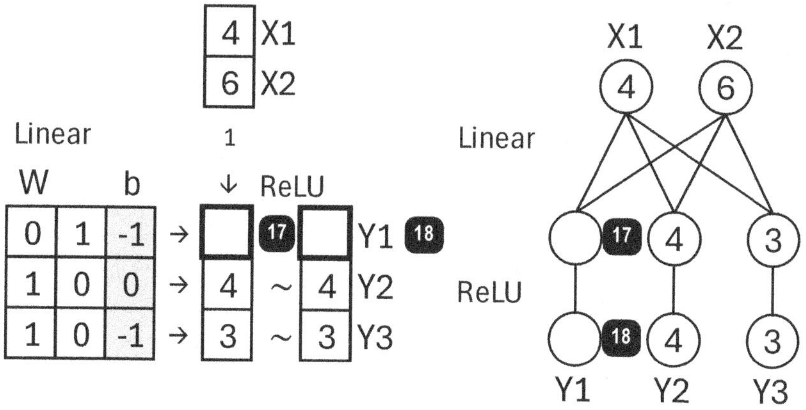

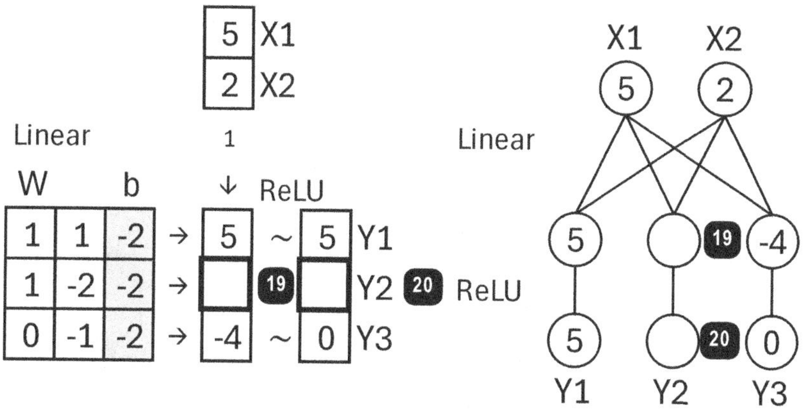

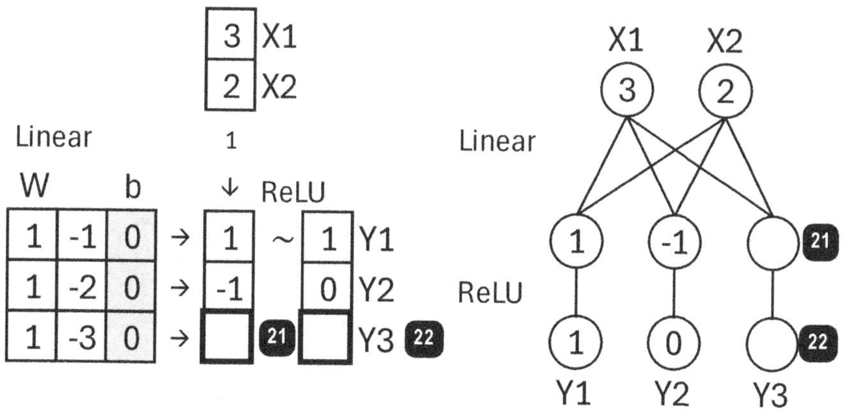

17	18	19	20	21	22
5	5	-1	0	-3	0

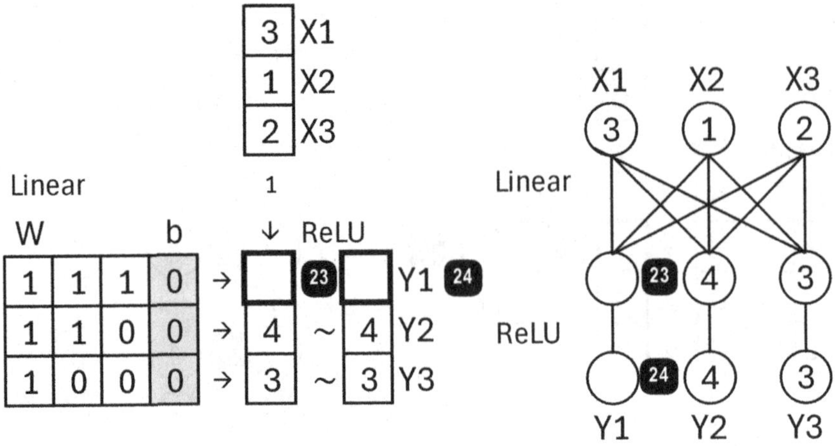

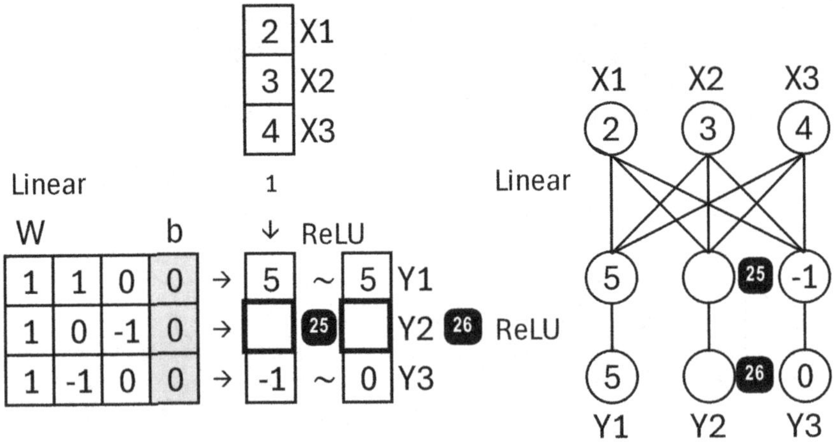

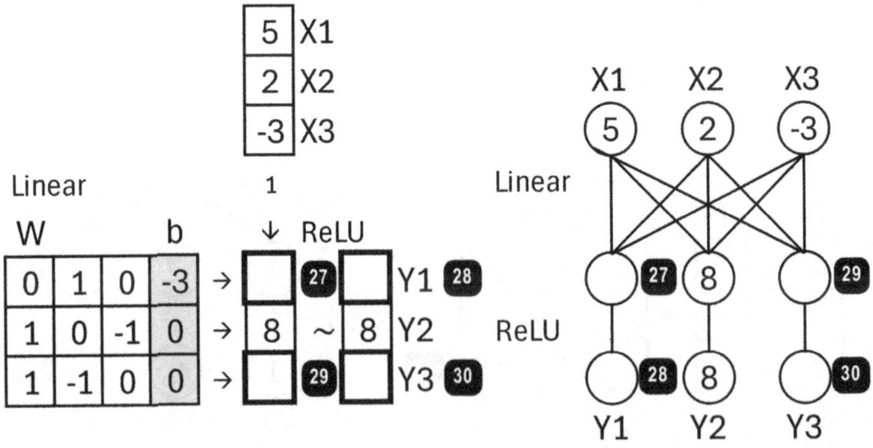

23	24	25	26	27	28	29	30
6	6	-2	0	-1	0	3	3

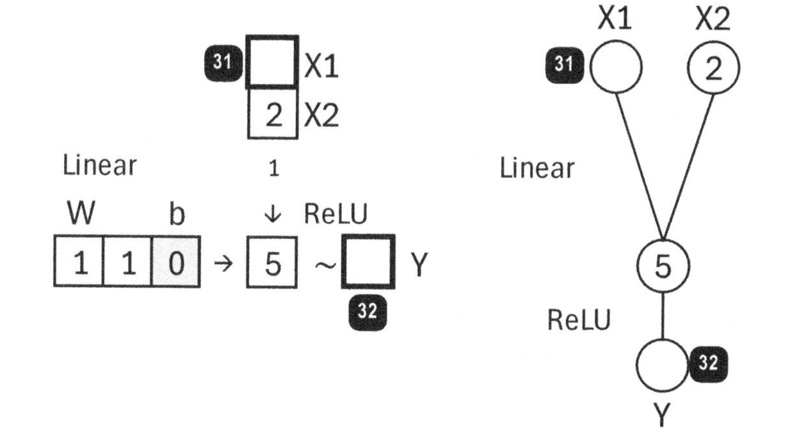

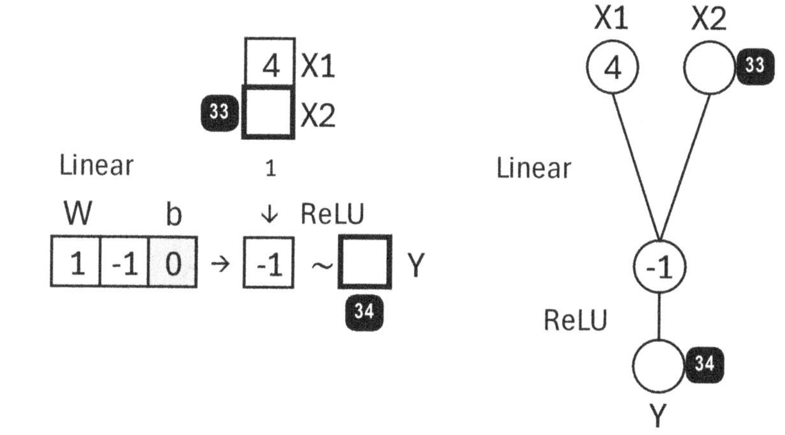

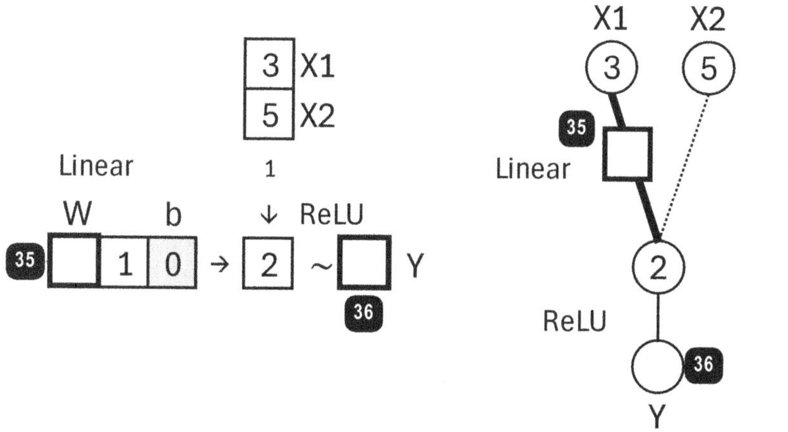

31	32	33	34	35	36
3	5	5	0	-1	2

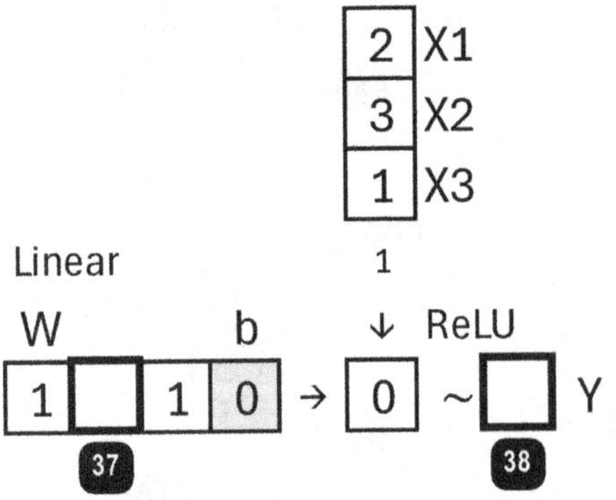

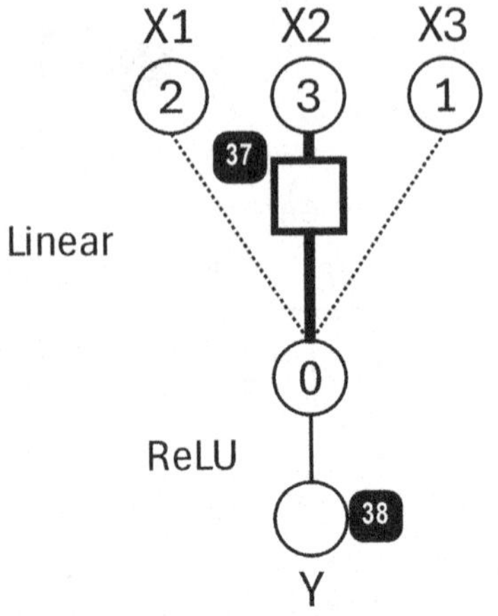

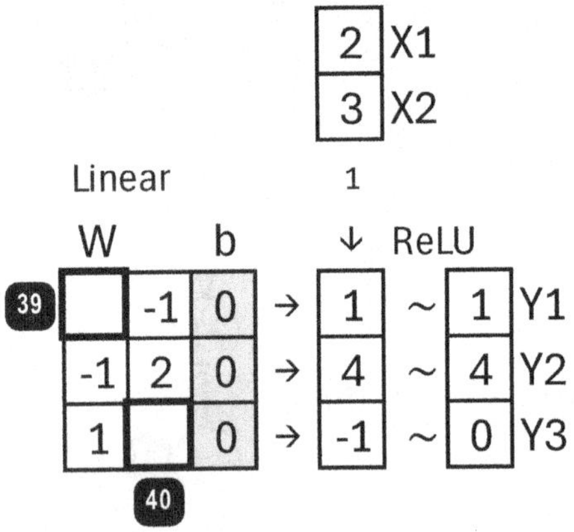

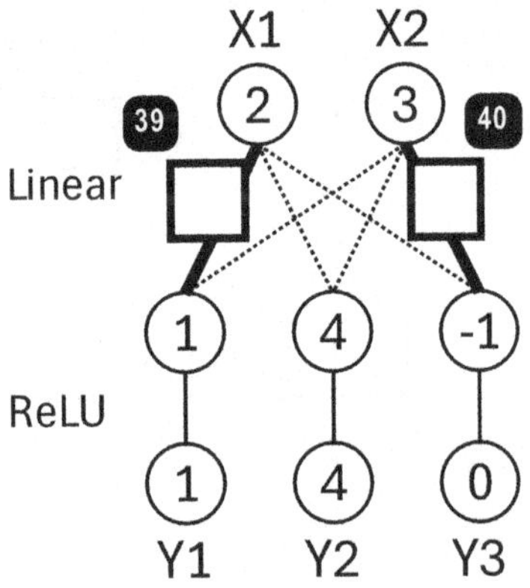

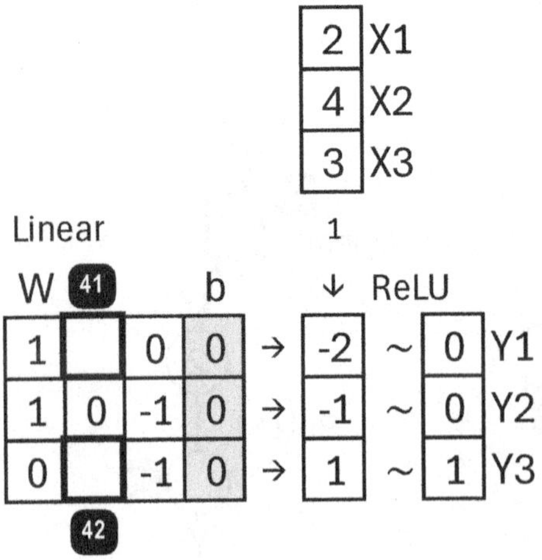

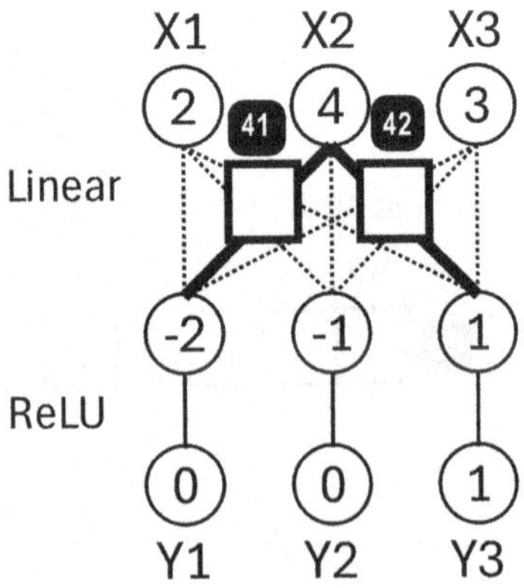

37	38	39	40	41	42
-1	0	2	-1	-1	1

(R) ReLU or (S) Sigmoid{3}

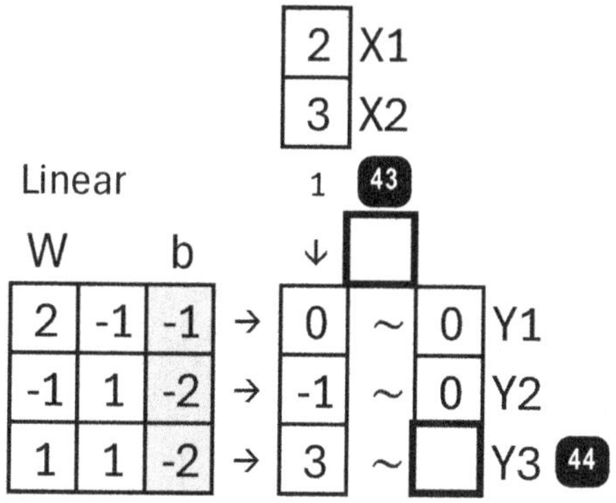

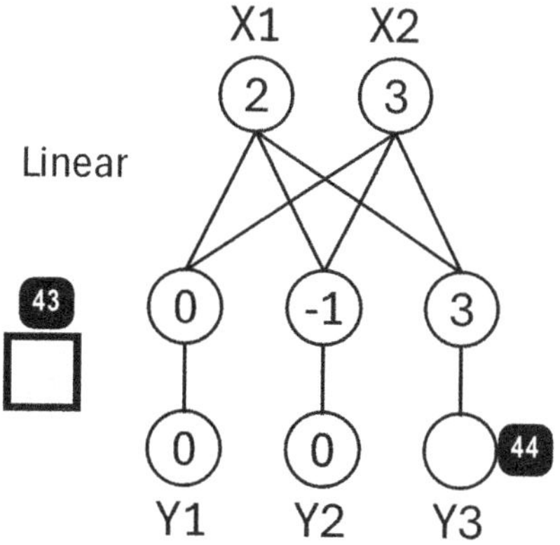

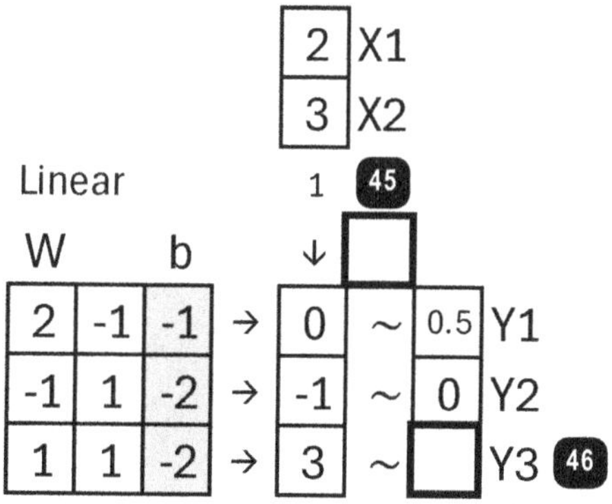

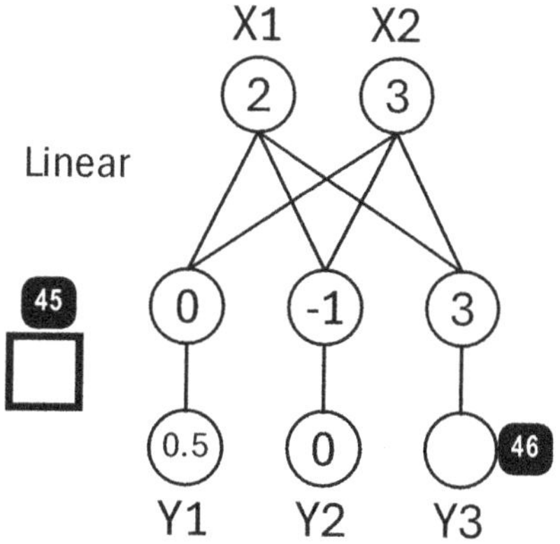

43	44	45	46
R	3	S	1

(R) ReLU or (S) Sigmoid{3}

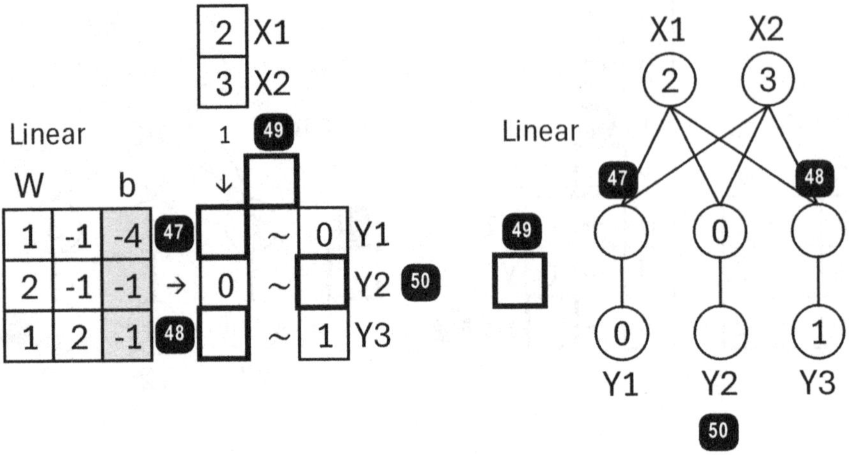

47	48	49	50
-5	7	S	0.5

Chapter 6

Batch

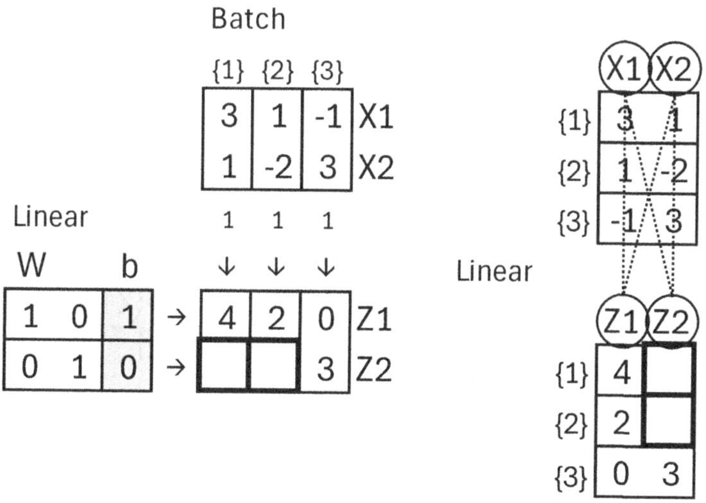

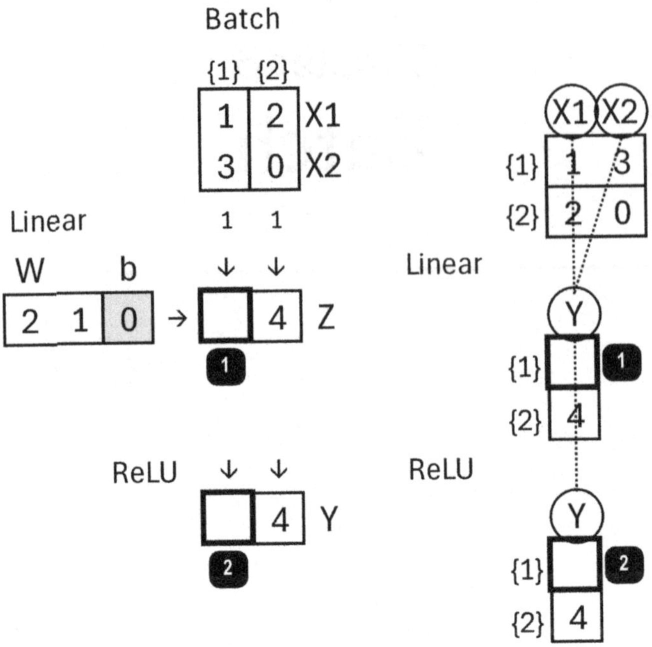
Batch
{1} {2}
1 2 X1
3 0 X2
Linear
1 1
W b
2 1 0
4 Z
1
ReLU
4 Y
2
X1 X2
{1} 1 3
{2} 2 0
Linear
Y
{1}
{2} 4
1
ReLU
Y
{1}
{2} 4
2

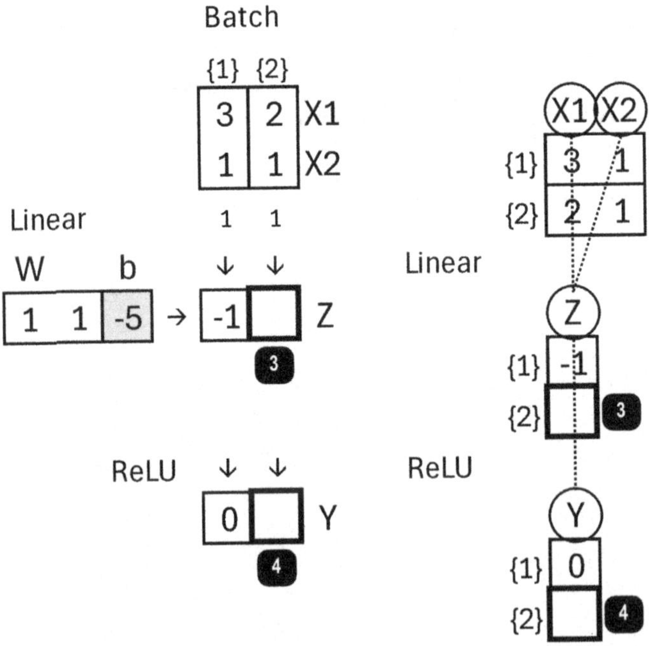
Batch
{1} {2}
3 2 X1
1 1 X2
Linear
1 1
W b
1 1 -5
-1 Z
3
ReLU
0 Y
4
X1 X2
{1} 3 1
{2} 2 1
Linear
Z
{1} -1
{2}
3
ReLU
Y
{1} 0
{2}
4

1	2	3	4
5	5	-2	0

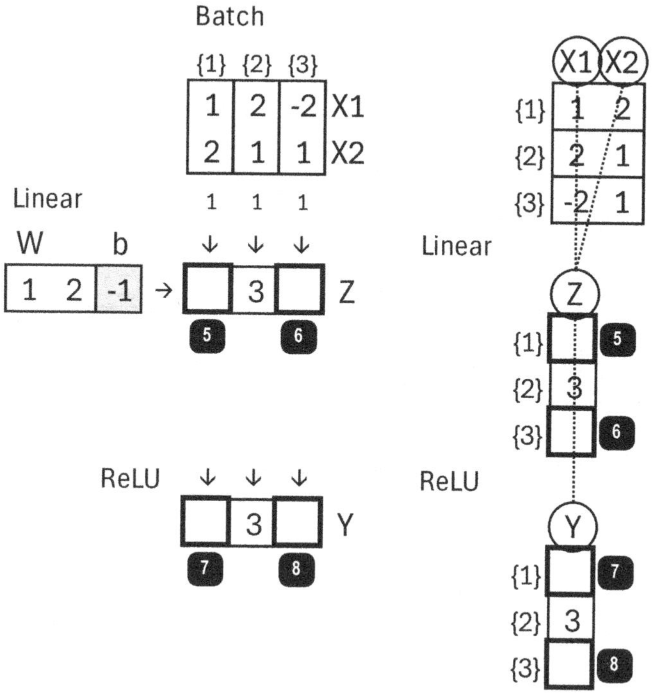

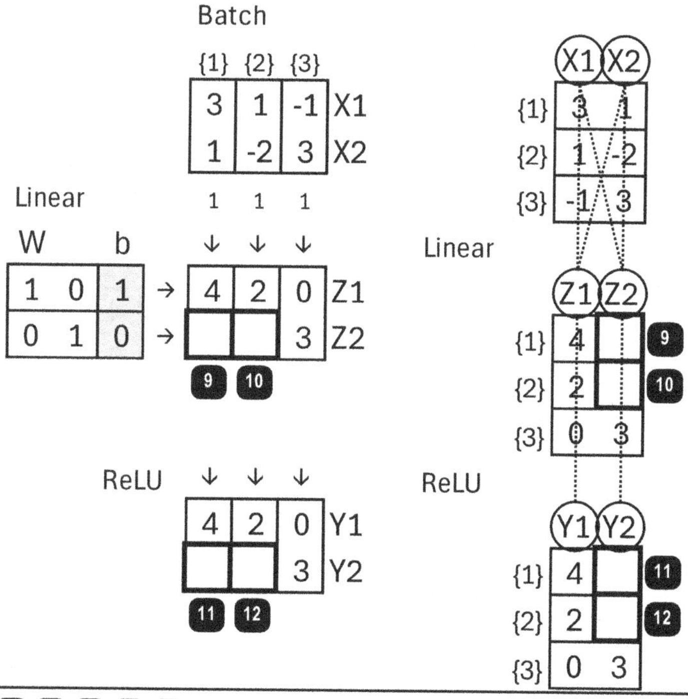

5	6	7	8	9	10	11	12
4	-1	4	0	1	-2	1	0

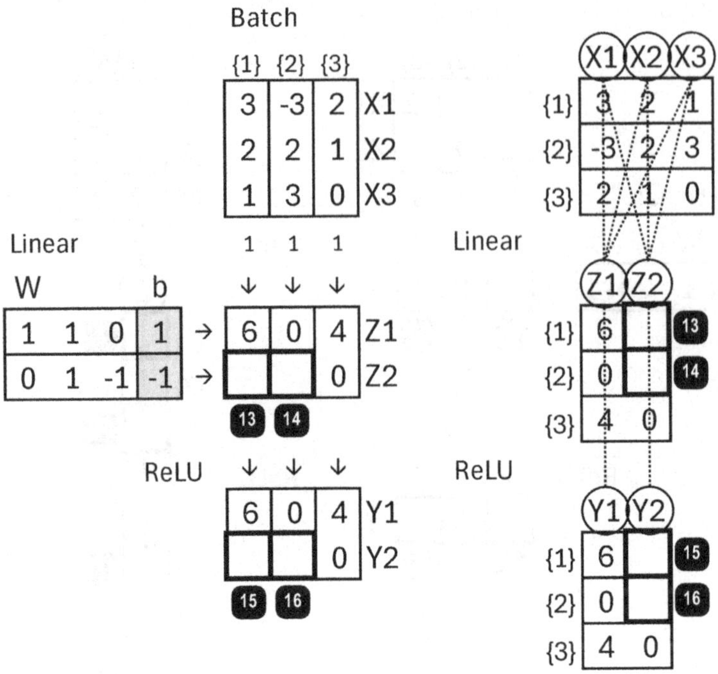

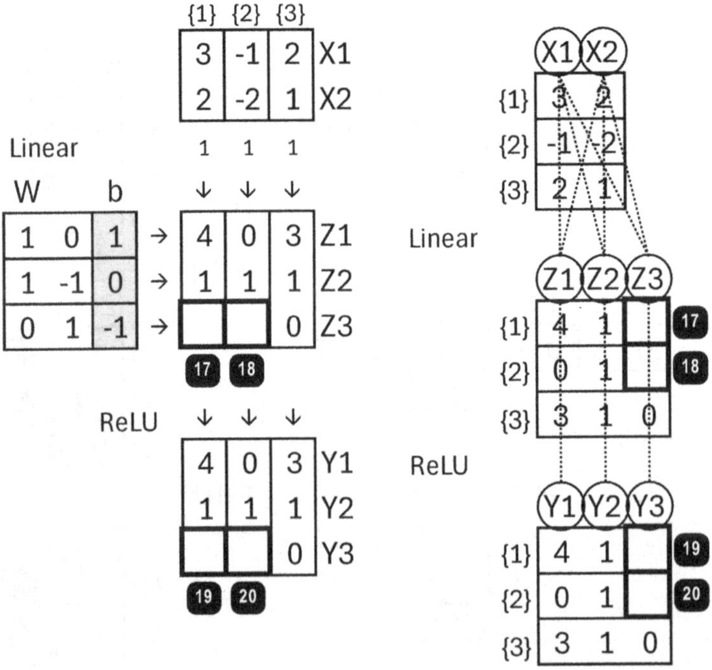

13	14	15	16	17	18	19	20
0	-2	0	0	1	-3	1	0

Batch

	{1}	{2}	{3}	{4}	{5}	{6}	
	3	2	3	2	3	0	X1
	1	3	1	-2	-1	2	X2
	1	1	1	1	1	1	

Linear

W		b
1	0	1
1	-1	0
0	1	-1

→

↓	↓	↓	↓	↓	↓	
4	3	4	3	4	1	Z1
2	-1	2	4	4	-2	Z2
0	2	0				Z3
			21	22	23	

ReLU

↓	↓	↓	↓	↓	↓	
4	3	4	3	4	1	Y1
2	0	2	4	4	0	Y2
0	2	0				Y3
			24	25	26	

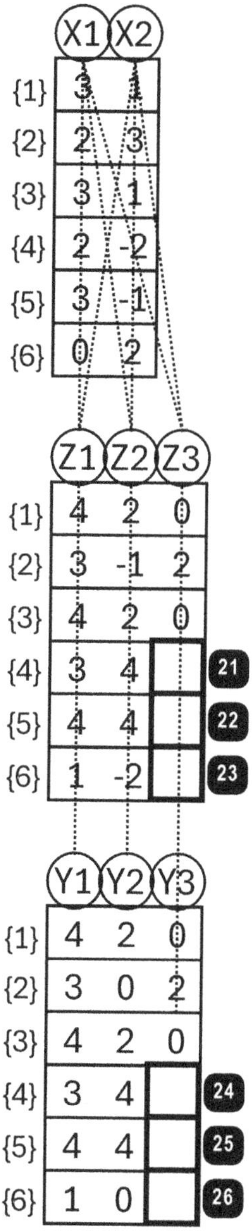

21	22	23	24	25	26
-3	-2	1	0	0	1

Batch Sum

	{1}	{2}	{3}	{4}		Σ
Y1	4	0	4	0	→	8
Y2	2	0	2	6	→	[27]
Y3	2	3	0	-3	→	2

	{1}	{2}	{3}	{4}	{5}	{6}	{7}	{8}	{9}	{10}		Σ
Y1	1	1	1	0	0	0	0	0	0	0	→	[28]
Y2	0	0	0	1	0	1	0	1	0	1	→	4
Y3	0	1	0	1	1	1	1	0	0	1	→	6

	{1}	{2}	{3}	{4}	{5}	{6}	{7}	{8}	{9}	{10}		Σ
Y1	1	1	1	0	0	0	0	0	0	[29]	→	2
Y2	0	0	2	1	0	1	0	1	0	1	→	6
Y3	0	1	0	1	[30]	1	1	0	0	1	→	3

Batch Mean

	{1}	{2}	{3}	{4}		μ
Y1	4	0	4	0	→	2
Y2	2	2	2	6	→	[31]
Y3	2	3	2	-3	→	[32]

	{1}	{2}	{3}	{4}	{5}	{6}	{7}	{8}	{9}	{10}	Σ
Y1	1	1	1	0	0	0	0	1	0	0	0.4
Y2	0	0	0	1	0	1	0	1	0	1	[33]
Y3	1	1	0	1	1	1	1	0	0	1	[34]

27	28	29	30	31	32	33	34
10	3	-1	-2	3	1	0.4	0.7

Batch Add

	{1}	{2}	{3}	{4}
Y1	4	0	4	0
Y2	2	2	2	6
Y3	2	3	2	-3

.+

3
1
2

=

{1}	{2}	{3}	{4}
7	3	7	. (35)
3	3	3	7
4	(36)	4	-1

	{1}	{2}	{3}	{4}
Y1	4	0	4	0
Y2	2	3	1	2
Y3	2	3	2	3

.+

(37)
-2
(38)

=

{1}	{2}	{3}	{4}
9	5	9	5
0	1	-1	0
5	6	5	6

Batch Subtract

	{1}	{2}	{3}	{4}
Y1	4	0	4	0
Y2	2	2	2	6
Y3	2	3	2	-3

.-

2
1
1

=

{1}	{2}	{3}	{4}
2	-2	2	-2
1	1	1	(39)
1	2	(40)	-4

	{1}	{2}	{3}	{4}
Y1	4	0	4	0
Y2	2	2	2	4
Y3	2	3	2	-3

.-

(41)
1
(42)

=

{1}	{2}	{3}	{4}
2	-2	2	-2
1	1	1	3
3	4	3	-2

35	36	37	38	39	40	41	42
3	5	5	3	5	1	2	-1

Batch Multiply

	{1}	{2}	{3}	{4}				{1}	{2}	{3}	{4}
Y1	4	0	4	0		2		8	0	8	0
Y2	2	1	2	-1	.*	3	=	6	3	6	(43)
Y3	2	3	2	-3		-1		-2	-3	(44)	3

	{1}	{2}	{3}	{4}				{1}	{2}	{3}	{4}
Y1	4	0	4	0		(45)		16	0	16	0
Y2	2	1	2	-1	.*	2	=	4	2	4	-2
Y3	2	1	3	-2		(46)		-6	-3	-9	6

Batch Center at Zero

	{1}	{2}	{3}	{4}		μ		{1}	{2}	{3}	{4}		μ'
Y1	2	0	4	2		(47)		0	-2	2	0	→	(48)
Y2	3	6	3	0	.-	3	=	0	3	0	-3	→	0
Y3	-1	1	-1	1		0		-1	1	-1	1	→	0

	{1}	{2}	{3}	{4}		μ		{1}	{2}	{3}	{4}		μ'
Y1	4	0	4	0		2		2	-2	2	-2	→	0
Y2	4	2	1	1	.-	2	=	2	0	-1	-1	→	0
Y3	-1	-4	1	0		(49)		0	-3	2	1	→	(50)

43	44	45	46	47	48	49	50
-3	-2	4	-3	2	0	-1	0

Chapter 7

Connection

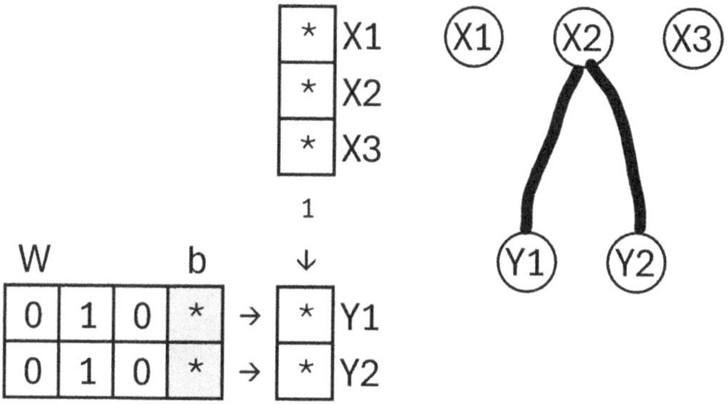

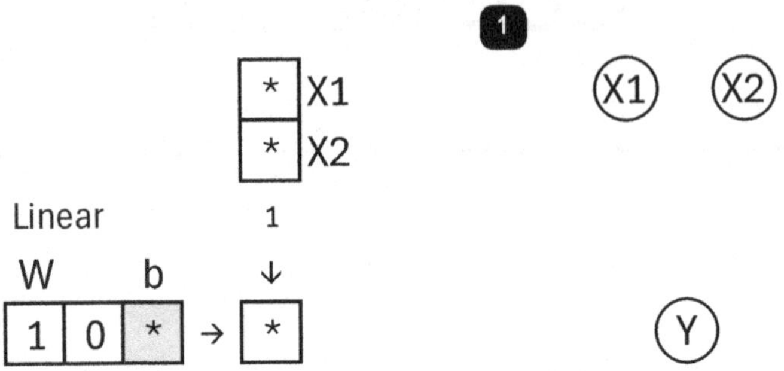
1
* X1
* X2
Linear
1
W b
1 0 *
*
X1 X2
Y

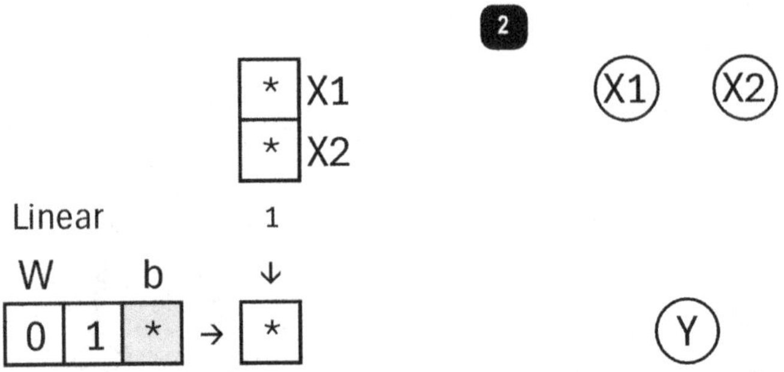
2
* X1
* X2
Linear
1
W b
0 1 *
*
X1 X2
Y

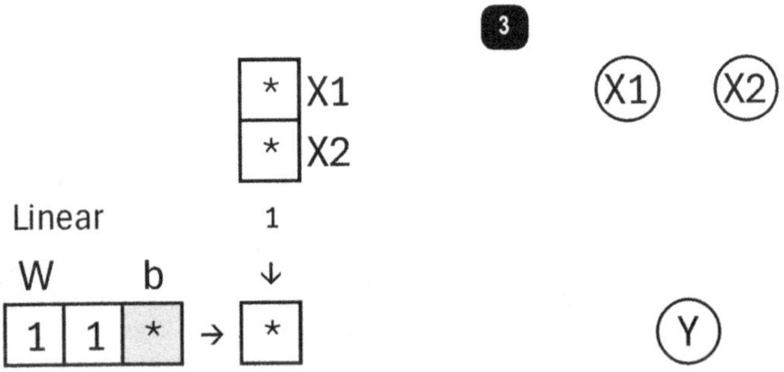
3
* X1
* X2
Linear
1
W b
1 1 *
*
X1 X2
Y

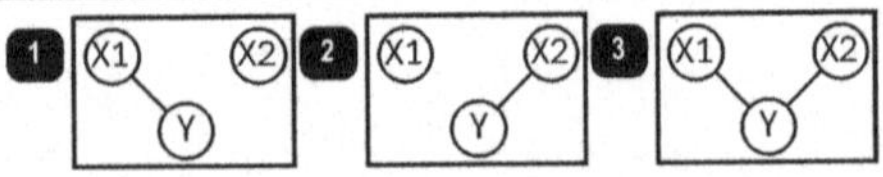
1 X1 X2 Y
2 X1 X2 Y
3 X1 X2 Y

4

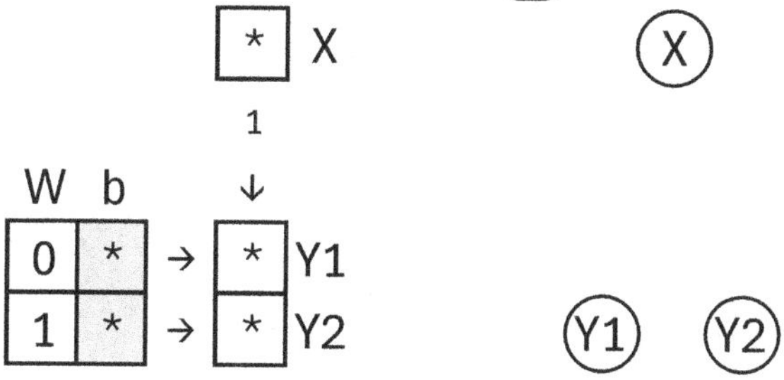

*
X
1
W
b
0
*
1
*
Y1
Y2
X
Y1
Y2

5

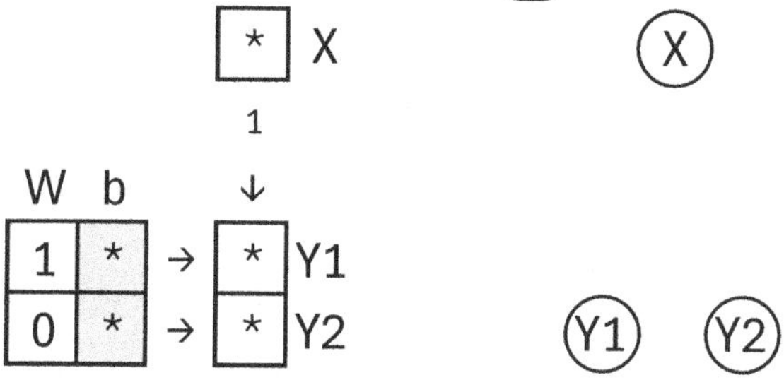

*
X
1
W
b
1
*
0
*
Y1
Y2
X
Y1
Y2

6

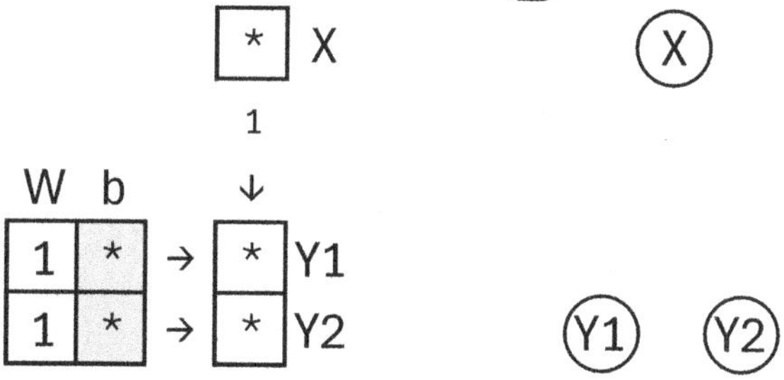

*
X
1
W
b
1
*
1
*
Y1
Y2
X
Y1
Y2

4

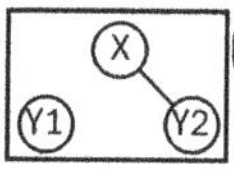

X
Y1
Y2

5

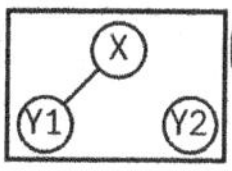

X
Y1
Y2

6

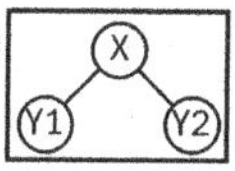

X
Y1
Y2

7

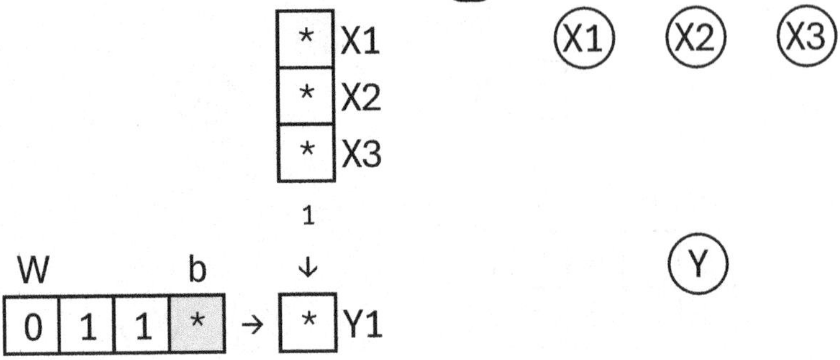

*
*
*
X1
X2
X3
1
W
b
0
1
1
*
*
Y1
X1
X2
X3
Y

8

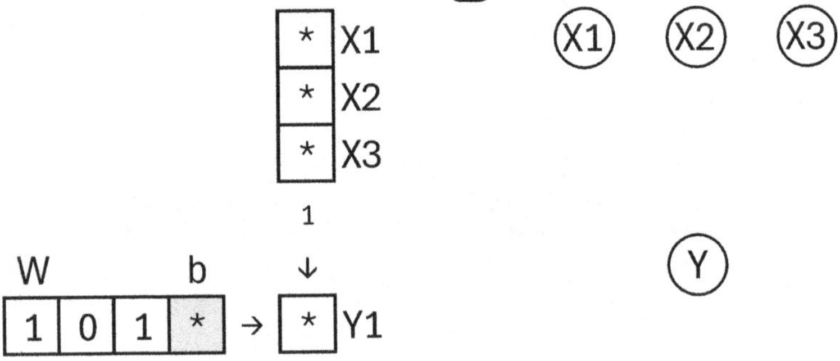

*
*
*
X1
X2
X3
1
W
b
1
0
1
*
*
Y1
X1
X2
X3
Y

9

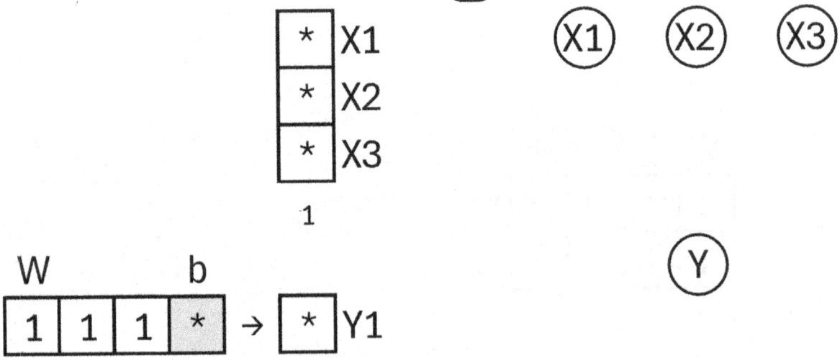

*
*
*
X1
X2
X3
1
W
b
1
1
1
*
*
Y1
X1
X2
X3
Y

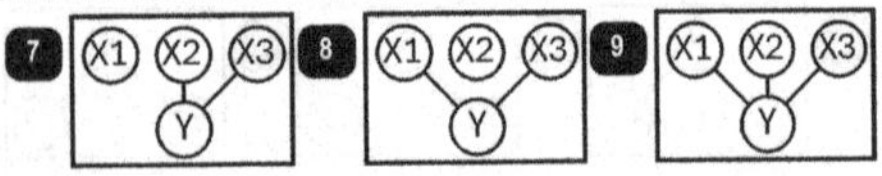

7
X1
X2
X3
Y
8
X1
X2
X3
Y
9
X1
X2
X3
Y

10

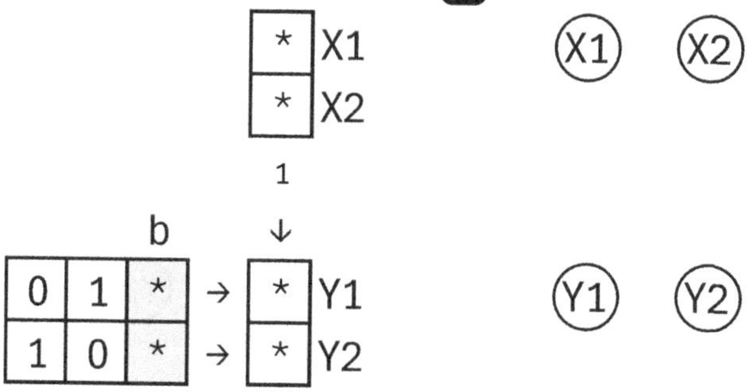
* X1
* X2
1
b
0 1 *
1 0 *
* Y1
* Y2
X1 X2
Y1 Y2

11

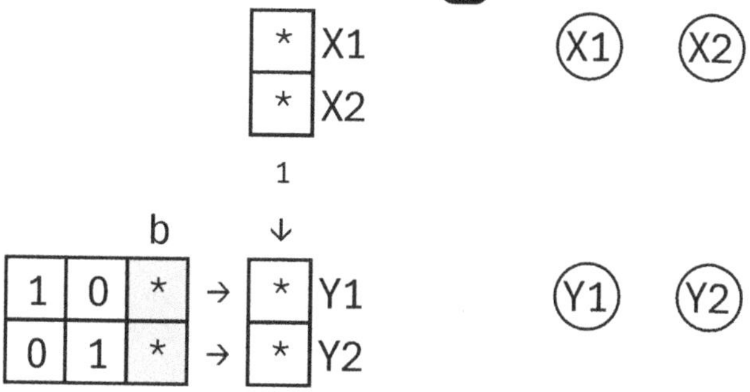
* X1
* X2
1
b
1 0 *
0 1 *
* Y1
* Y2
X1 X2
Y1 Y2

12

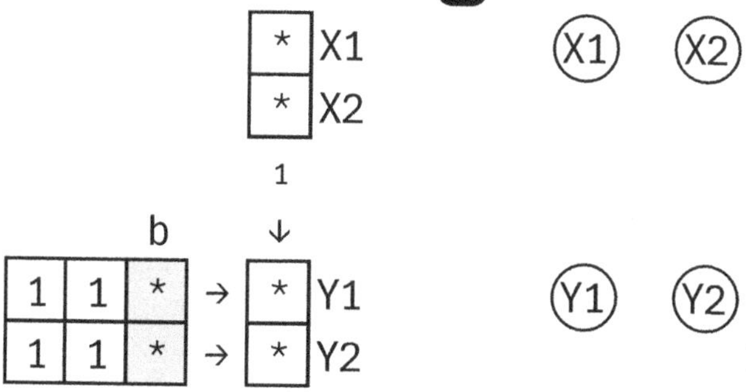
* X1
* X2
1
b
1 1 *
1 1 *
* Y1
* Y2
X1 X2
Y1 Y2

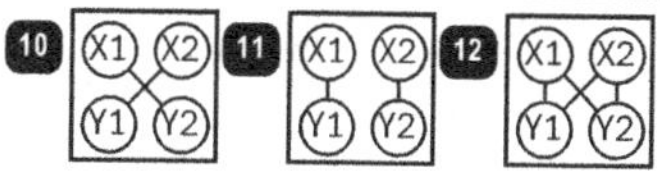
10
11
12

13

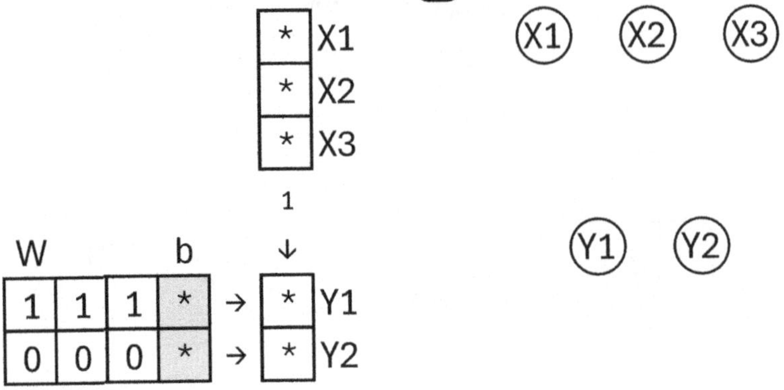

X1
X2
X3
1
W
b
1 1 1 *
0 0 0 *
Y1
Y2

14

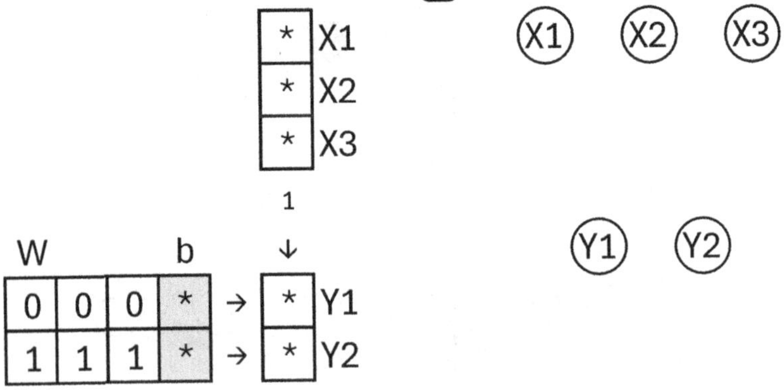

X1
X2
X3
1
W
b
0 0 0 *
1 1 1 *
Y1
Y2

15

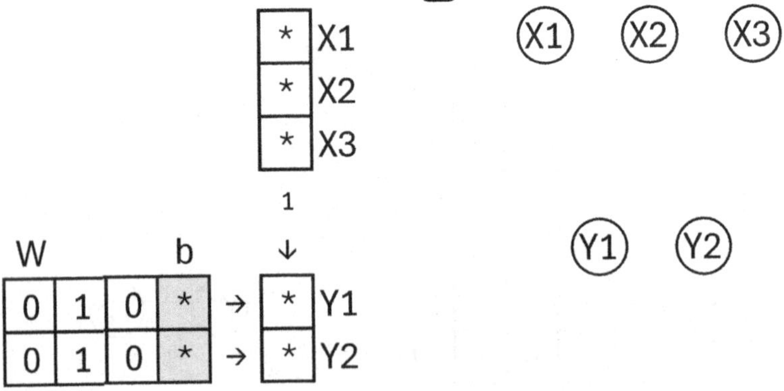

X1
X2
X3
1
W
b
0 1 0 *
0 1 0 *
Y1
Y2

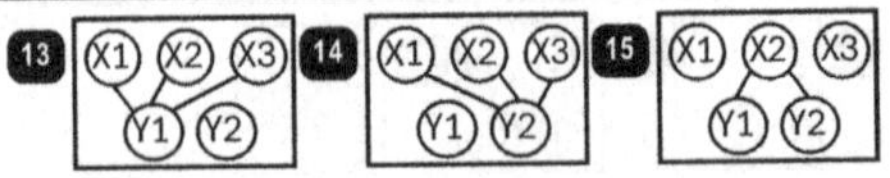

13
14
15
X1
X2
X3
Y1
Y2

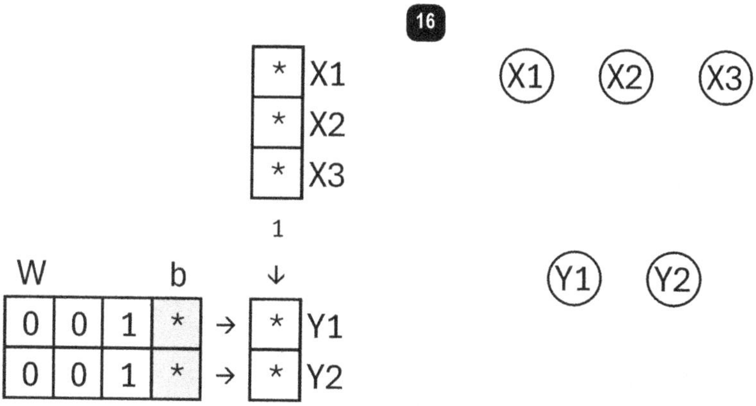
16
* X1
* X2
* X3
1
W
b
0 0 1 * → * Y1
0 0 1 * → * Y2
X1 X2 X3
Y1 Y2

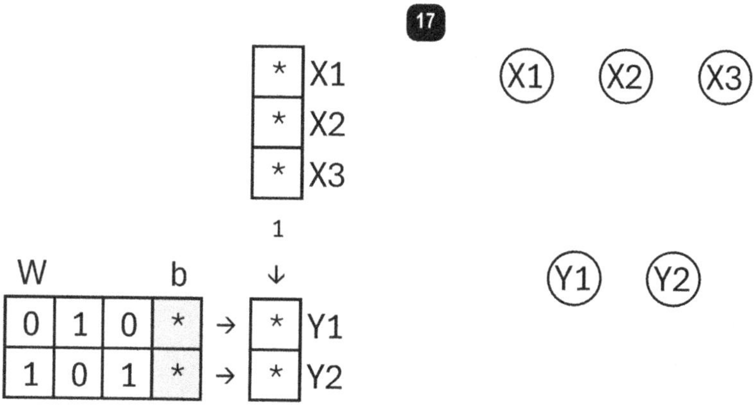
17
* X1
* X2
* X3
1
W
b
0 1 0 * → * Y1
1 0 1 * → * Y2
X1 X2 X3
Y1 Y2

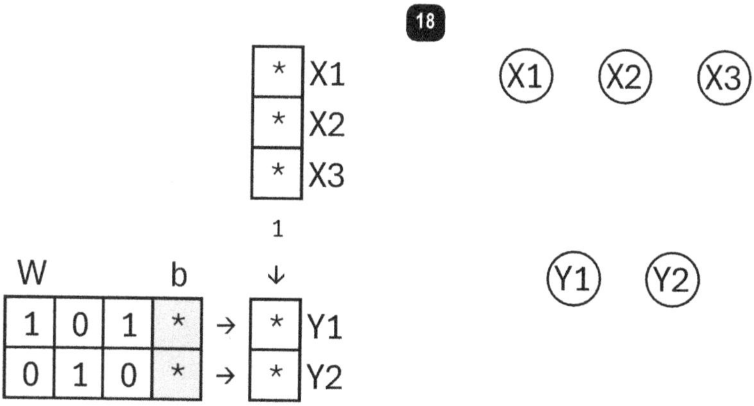
18
* X1
* X2
* X3
1
W
b
1 0 1 * → * Y1
0 1 0 * → * Y2
X1 X2 X3
Y1 Y2

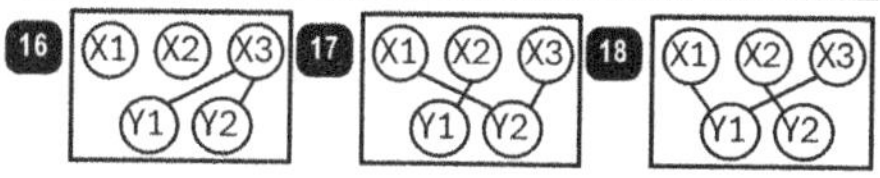
16
X1 X2 X3
Y1 Y2
17
X1 X2 X3
Y1 Y2
18
X1 X2 X3
Y1 Y2

19

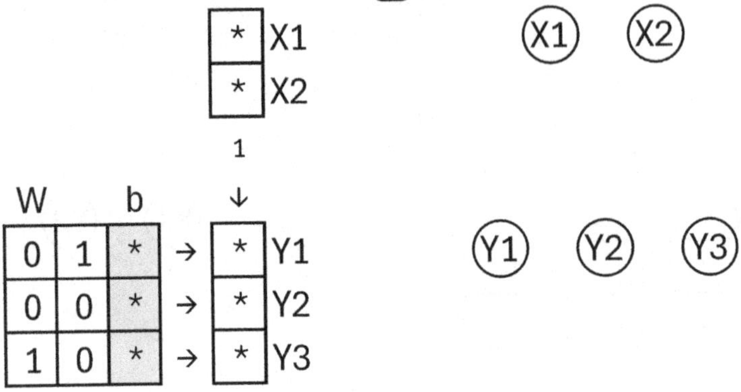

20

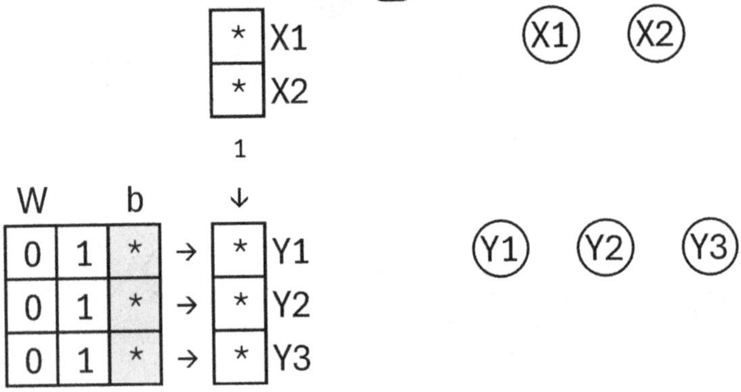

21

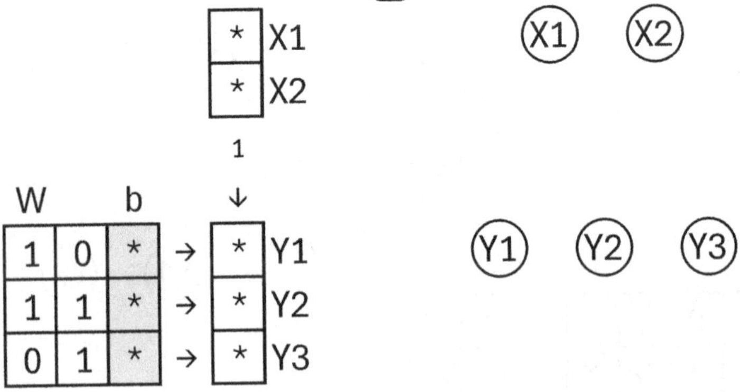

19

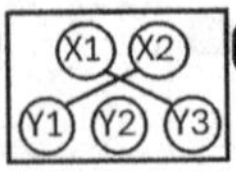

20

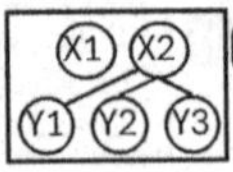

21

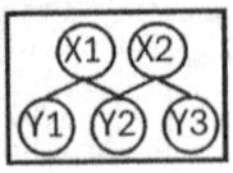

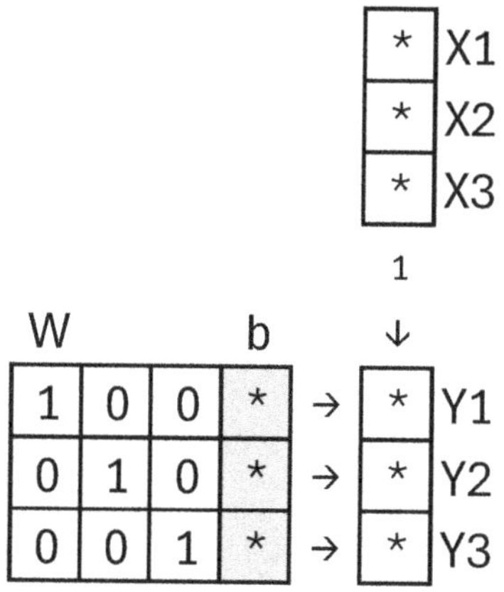

22

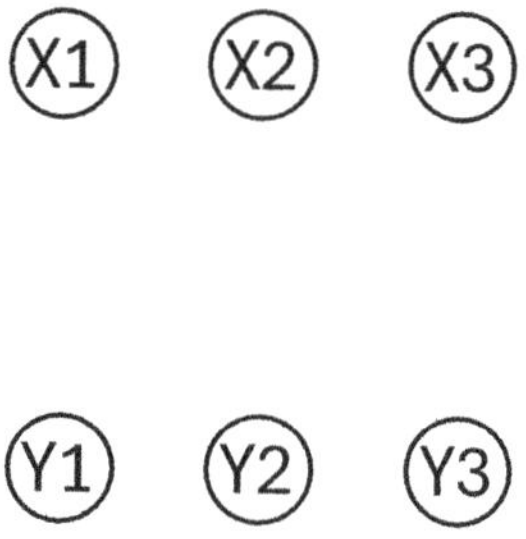

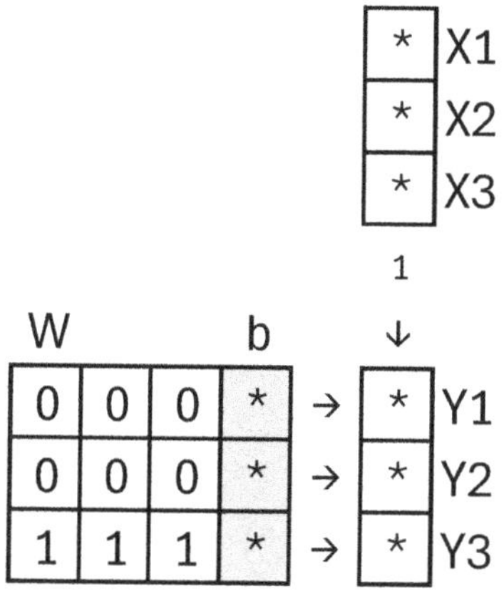

23

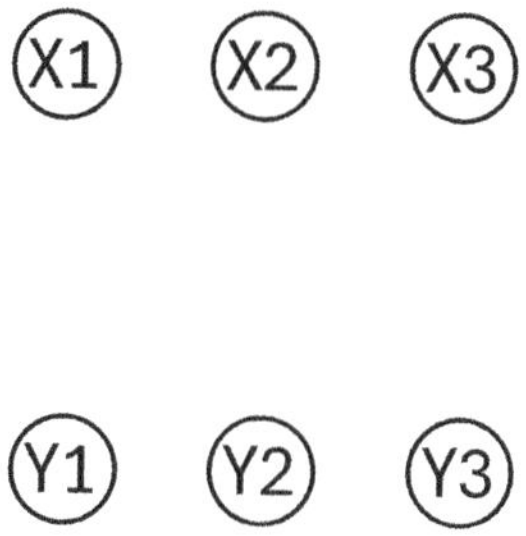

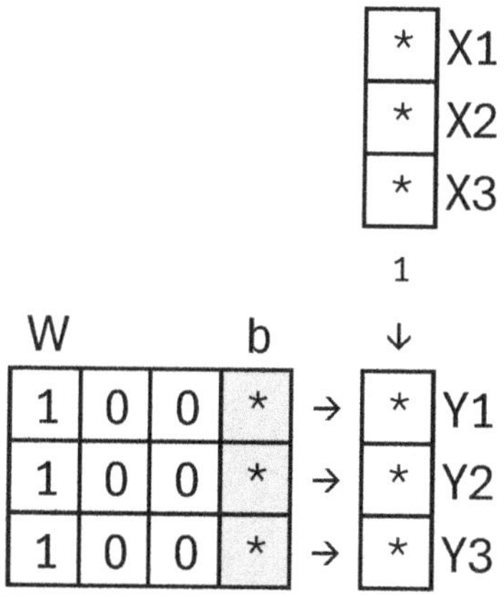

24

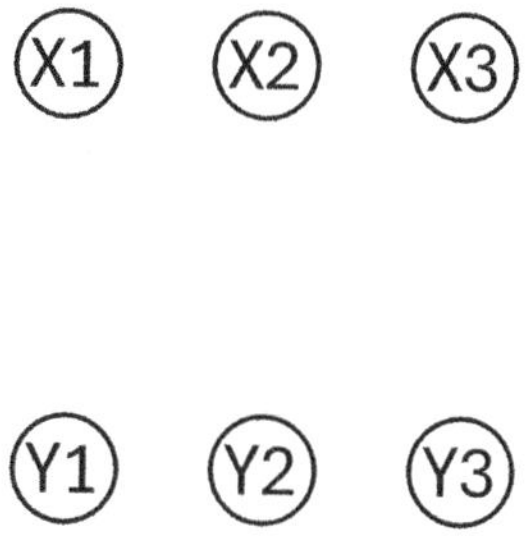

22
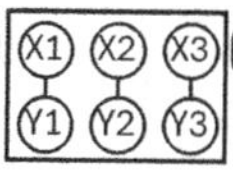
23
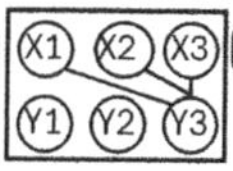
24
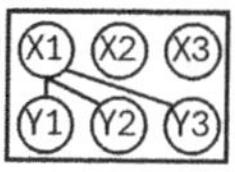

25

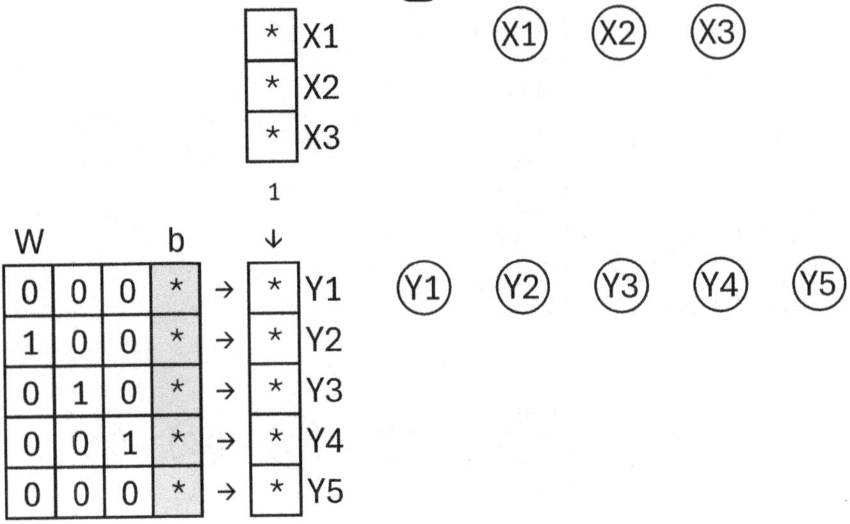
* X1
* X2
* X3
1
W
b
0 0 0 * → * Y1
1 0 0 * → * Y2
0 1 0 * → * Y3
0 0 1 * → * Y4
0 0 0 * → * Y5
X1 X2 X3
Y1 Y2 Y3 Y4 Y5

26

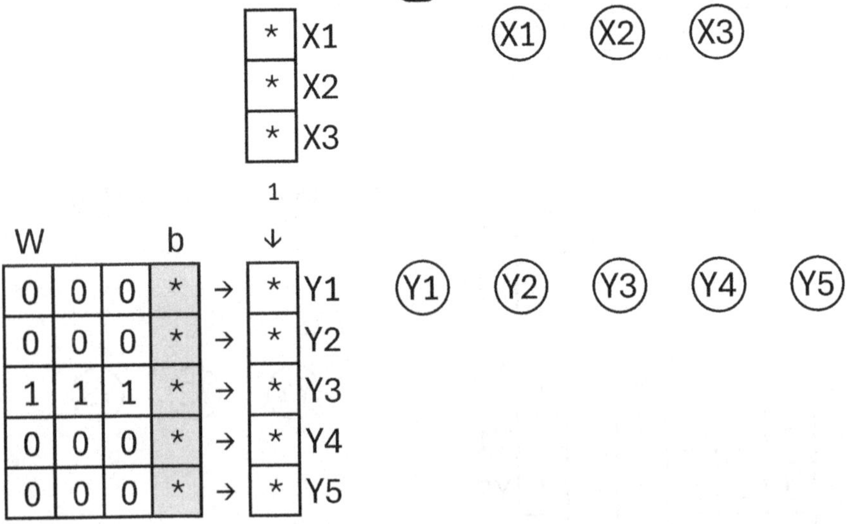
* X1
* X2
* X3
1
W
b
0 0 0 * → * Y1
0 0 0 * → * Y2
1 1 1 * → * Y3
0 0 0 * → * Y4
0 0 0 * → * Y5
X1 X2 X3
Y1 Y2 Y3 Y4 Y5

25

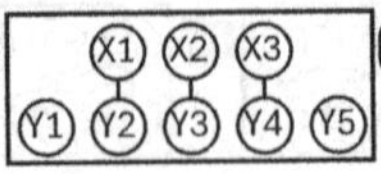
X1 X2 X3
Y1 Y2 Y3 Y4 Y5

26

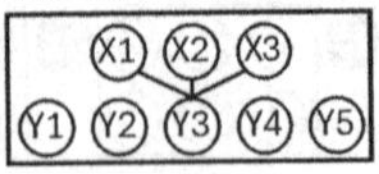
X1 X2 X3
Y1 Y2 Y3 Y4 Y5

27

*	X1
*	X2
*	X3

1

↓

W			b			
0	1	0	*	→	*	Y1
0	1	0	*	→	*	Y2
0	1	0	*	→	*	Y3
0	1	0	*	→	*	Y4
0	1	0	*	→	*	Y5

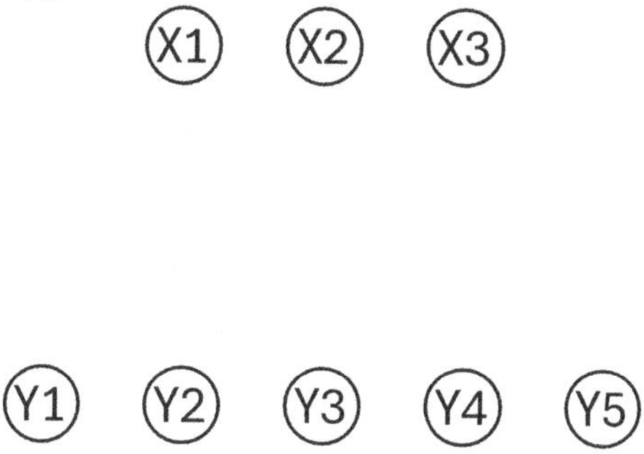

28

*	X1
*	X2
*	X3

1

↓

W			b			
0	0	0	*	→	*	Y1
1	1	0	*	→	*	Y2
0	0	0	*	→	*	Y3
0	0	0	*	→	*	Y4
0	1	1	*	→	*	Y5

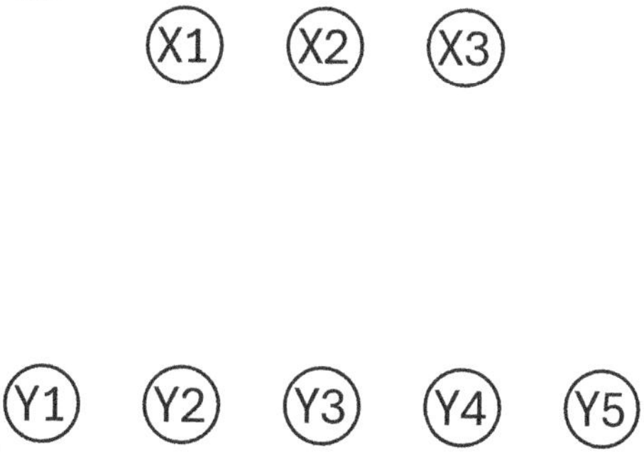

27
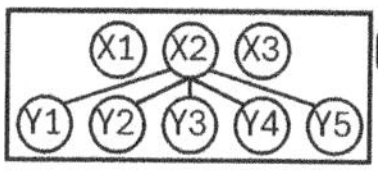

28
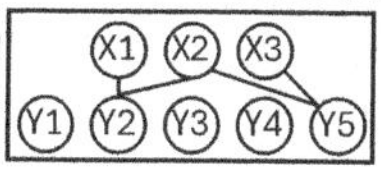

29

*	X1
*	X2
*	X3
*	X4
*	X5

1

↓

W					b			
1	1	0	0	0	*	→	*	Y1
0	1	1	0	0	*	→	*	Y2
0	0	1	1	0	*	→	*	Y3
0	0	0	1	1	*	→	*	Y4
0	0	0	0	1	*	→	*	Y5

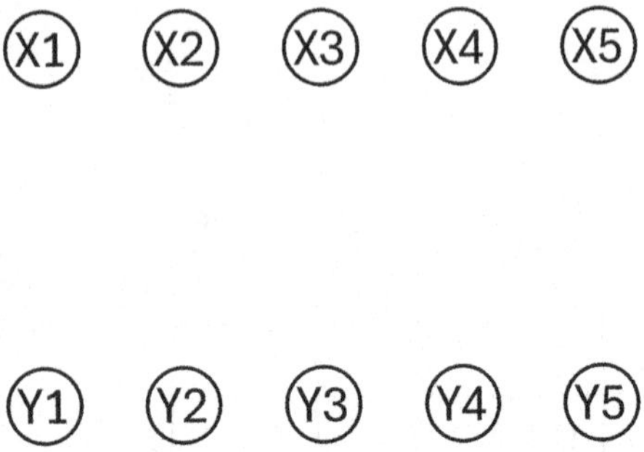

30

*	X1
*	X2
*	X3
*	X4
*	X5

1

↓

W					b			
1	1	0	0	0	*	→	*	Y1
1	1	0	0	0	*	→	*	Y2
0	0	1	0	0	*	→	*	Y3
0	o	0	1	1	*	→	*	Y4
0	0	0	1	1	*	→	*	Y5

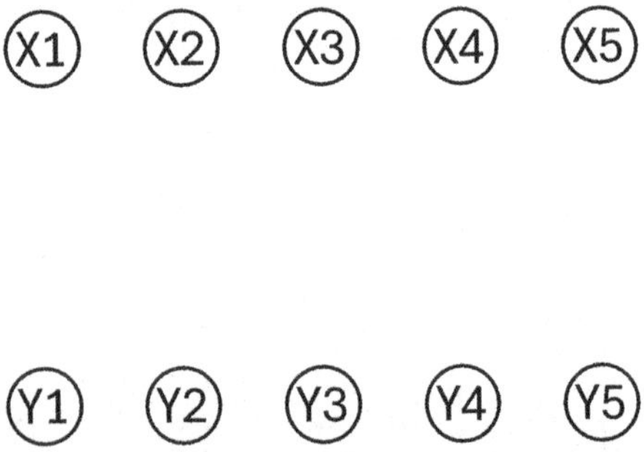

29

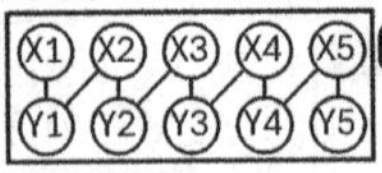

30

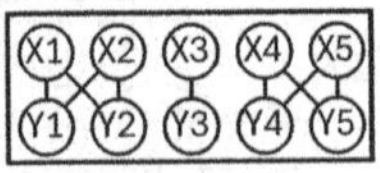

Chapter 8

Hidden Layer

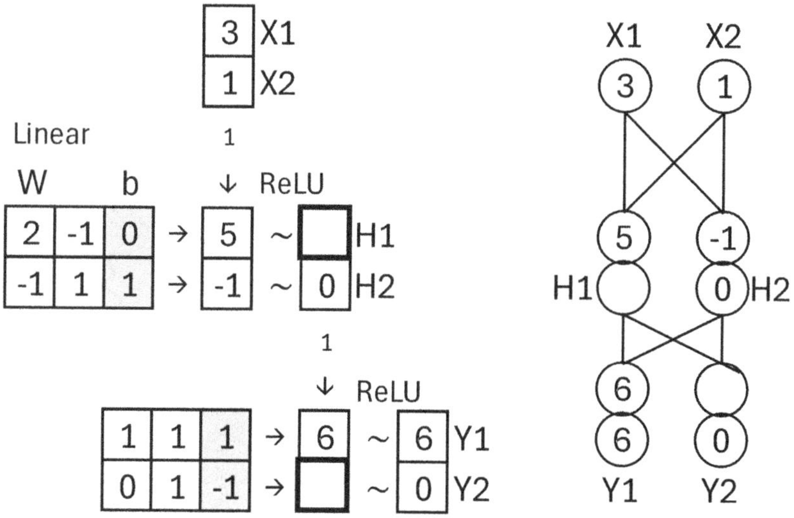

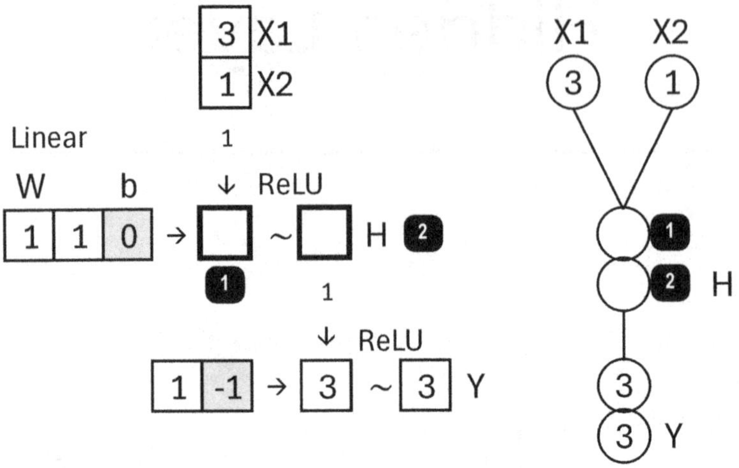
3 X1
1 X2
Linear
1
W b
ReLU
1 1 0
H 2
1
1
ReLU
1 -1
3
3 Y
X1 X2
3 1
1
2 H
3
3 Y

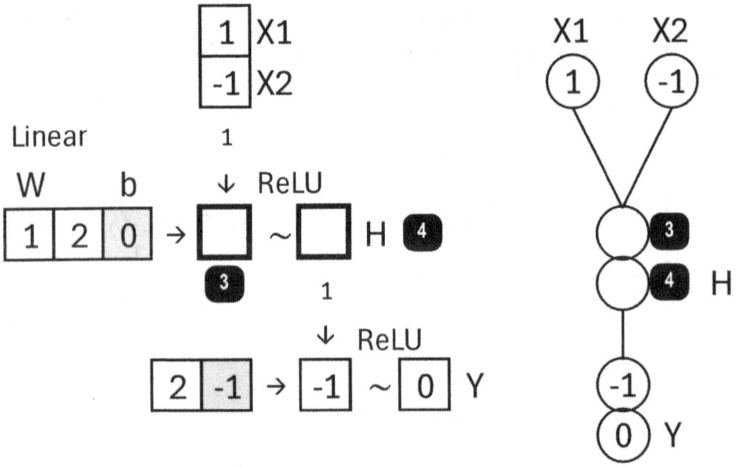
1 X1
-1 X2
Linear
1
W b
ReLU
1 2 0
H 4
3
1
ReLU
2 -1
-1
0 Y
X1 X2
1 -1
3
4 H
-1
0 Y

1	2	3	4
4	4	-1	0

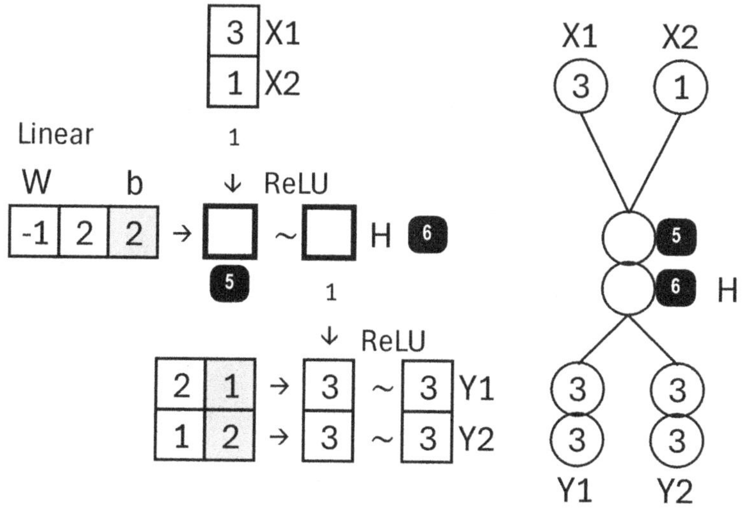
3 X1
1 X2
Linear
1
W b
↓ ReLU
-1 2 2 → ~ H 6
5
1
↓ ReLU
2 1 → 3 ~ 3 Y1
1 2 → 3 ~ 3 Y2
X1 X2
3 1
5
6 H
3 3
3 3
Y1 Y2

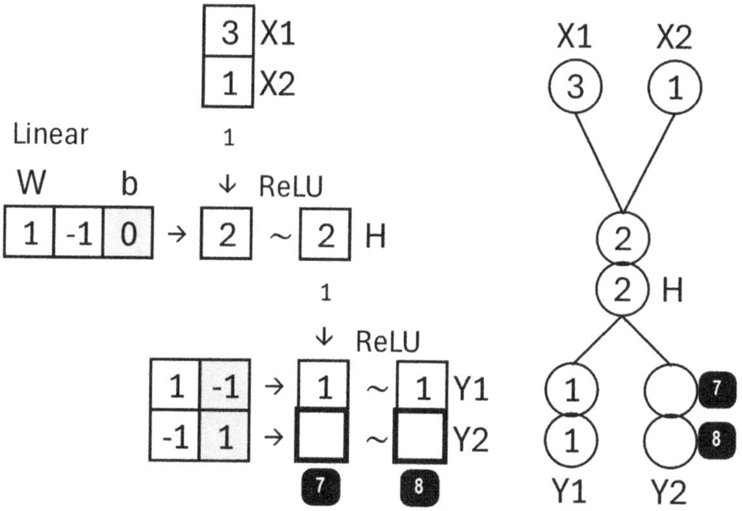
3 X1
1 X2
Linear
1
W b
↓ ReLU
1 -1 0 → 2 ~ 2 H
1
↓ ReLU
1 -1 → 1 ~ 1 Y1
-1 1 → ~ Y2
7 8
X1 X2
3 1
2
2 H
1 7
1 8
Y1 Y2

5	6	7	8
1	1	-1	0

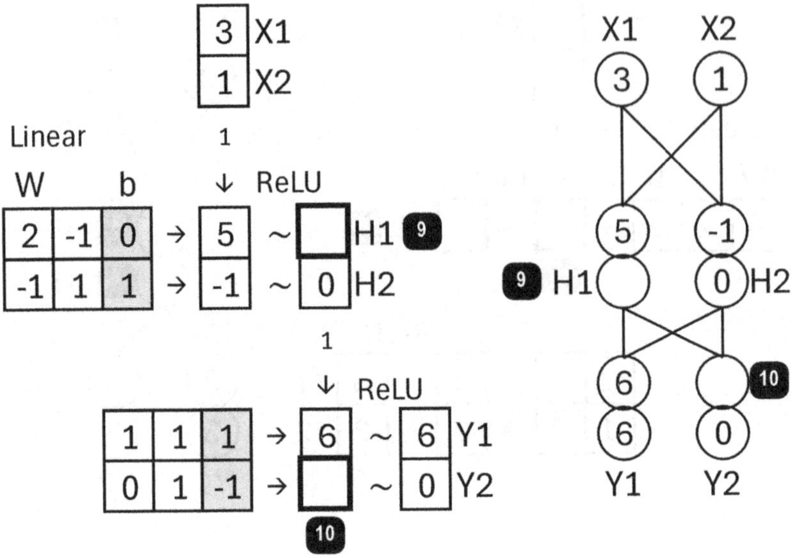

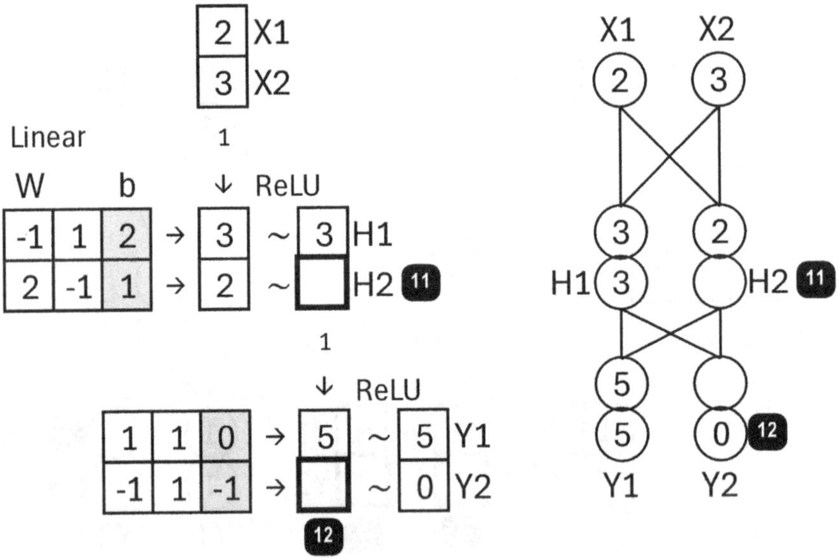

9	10	11	12
5	-1	2	-2

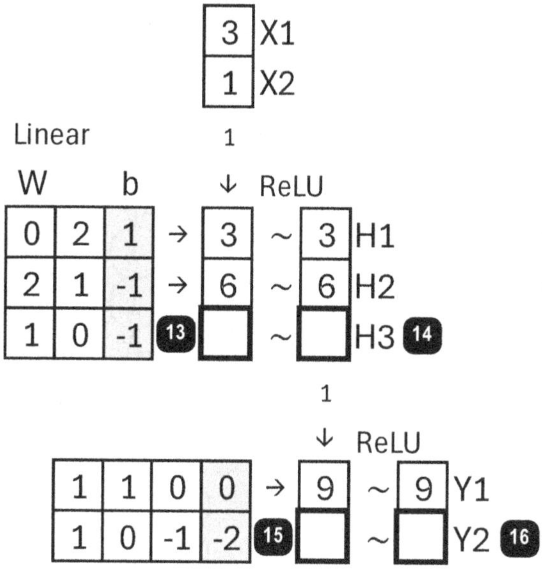

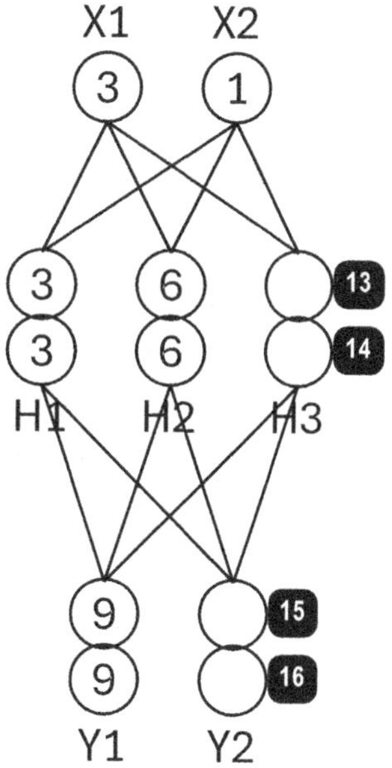

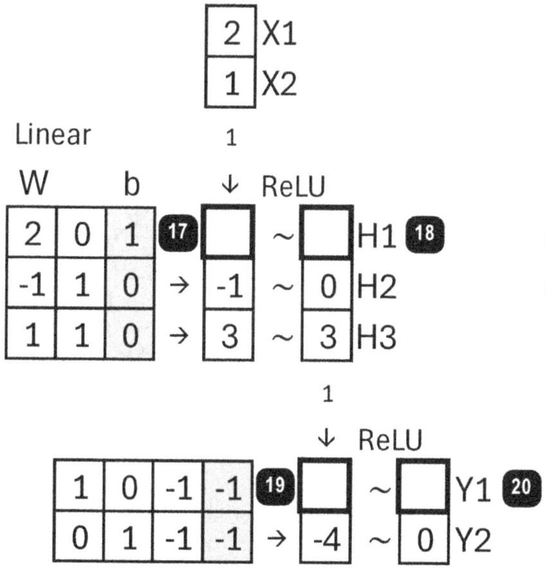

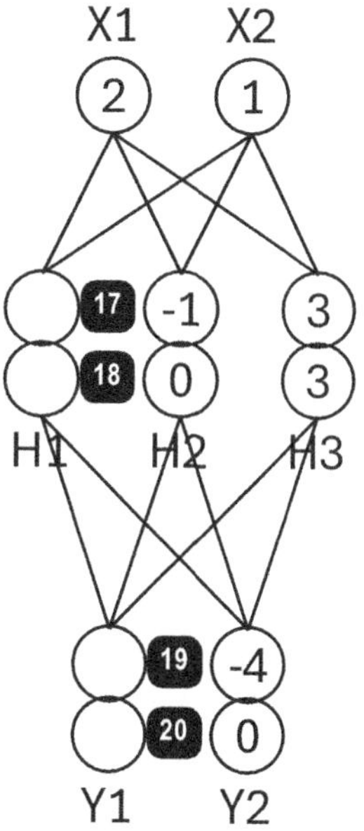

13	14	15	16	17	18	19	20
2	2	-1	0	5	5	1	1

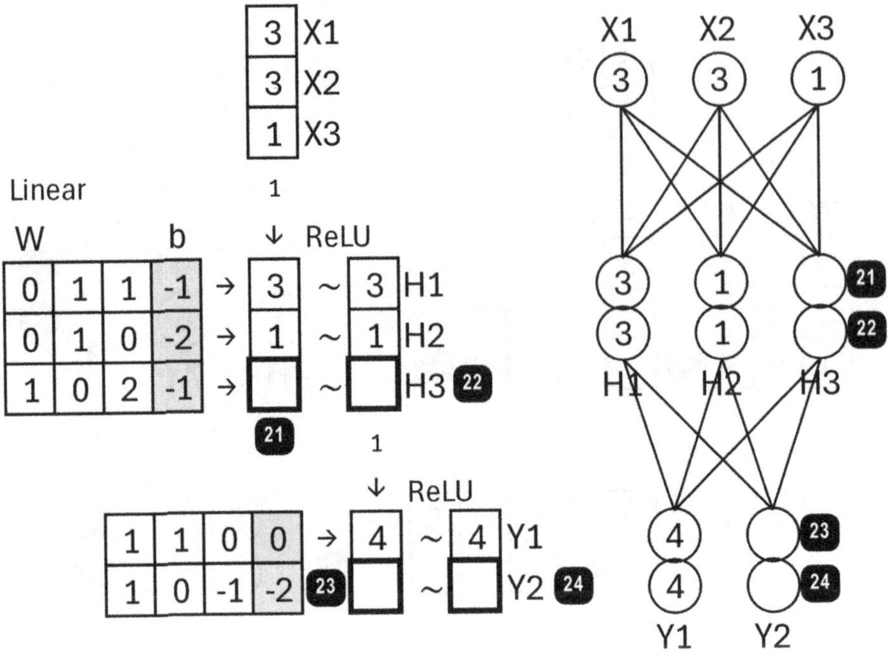

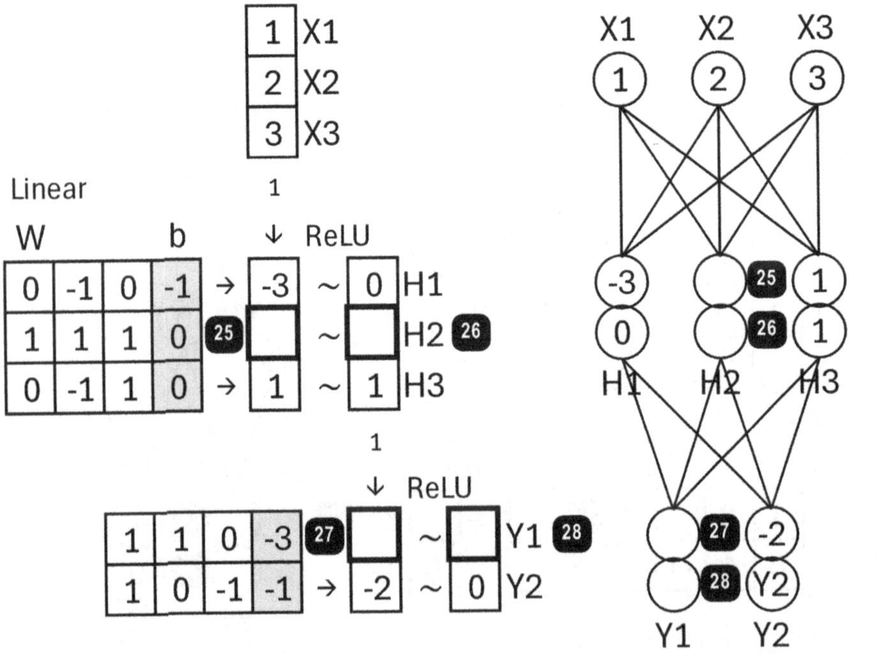

21	22	23	24	25	26	27	28
4	4	-3	0	6	6	3	3

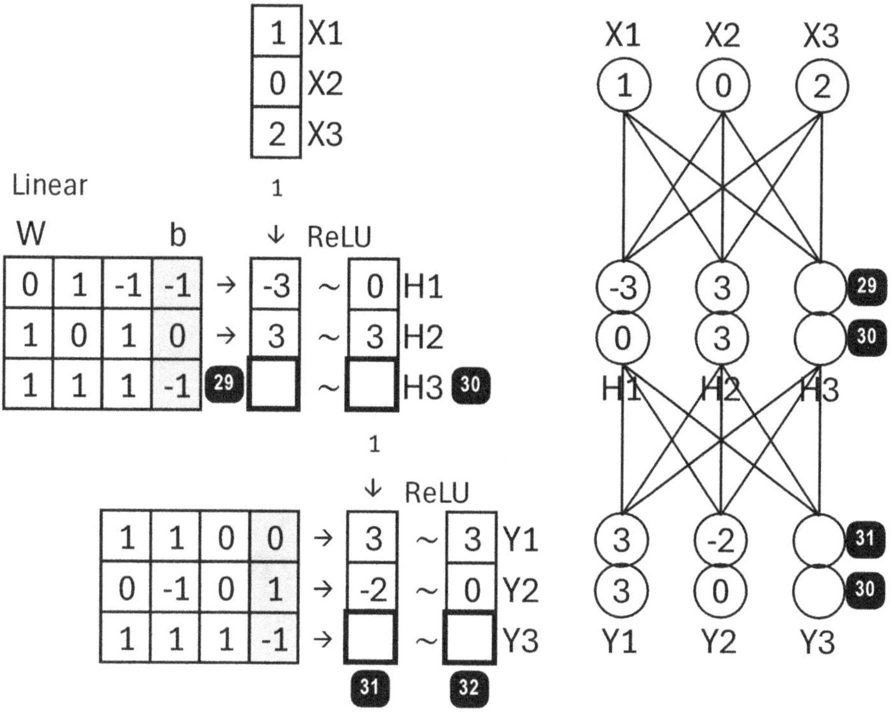

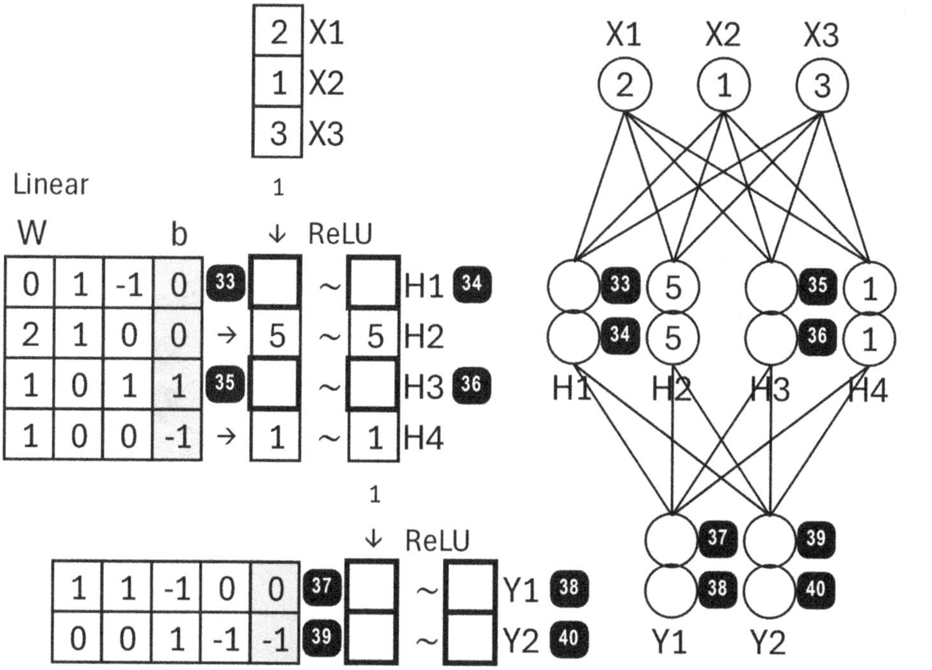

29	30	31	32	33	34	35	36	37	38	39	40
2	2	4	4	-2	0	6	6	-1	0	4	4

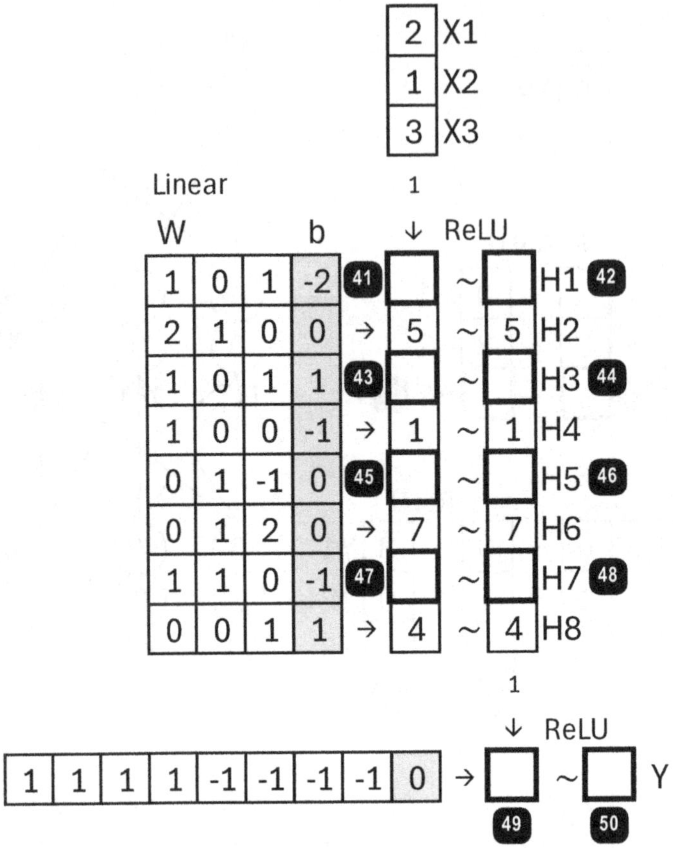

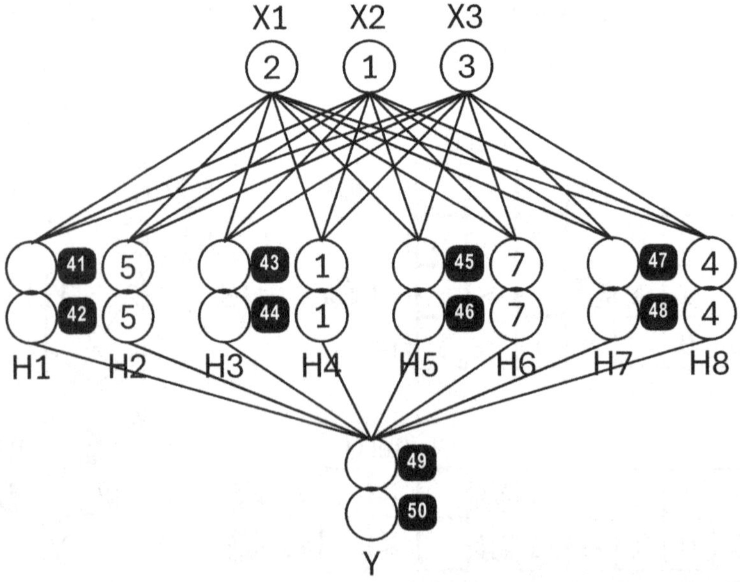

41	42	43	44	45	46	47	48	49	50
3	3	6	6	-2	0	2	2	2	2

Chapter 9
Deep

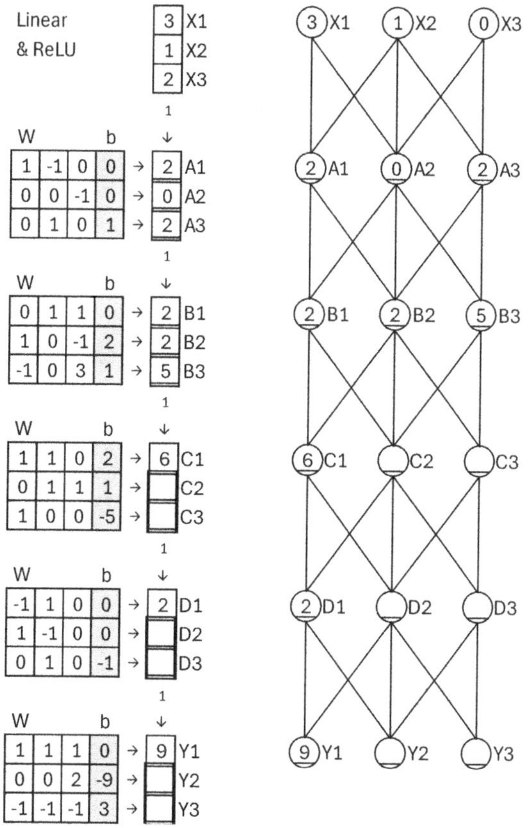

Linear

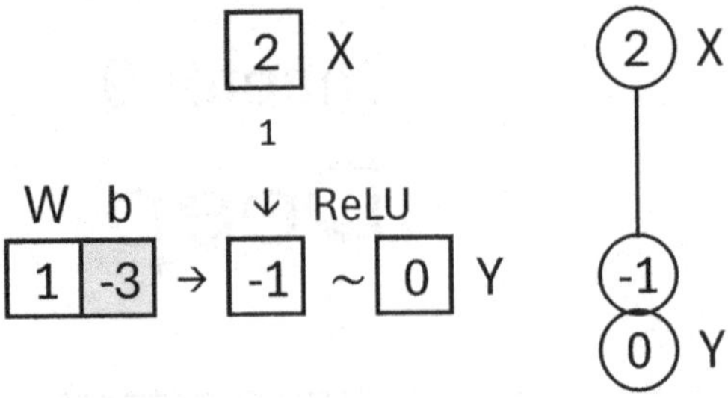

Linear
& ReLU

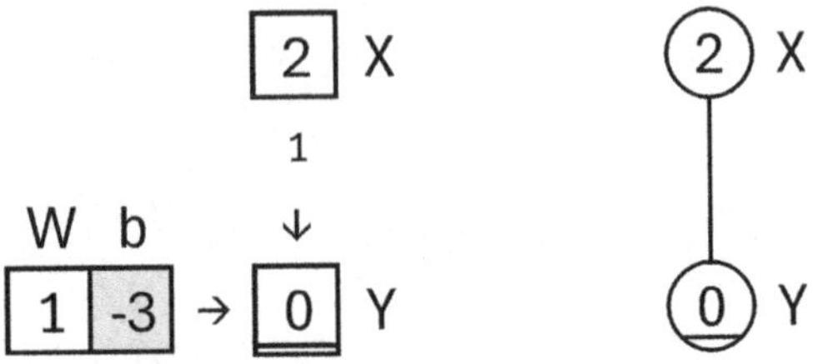

Linear
& ReLU

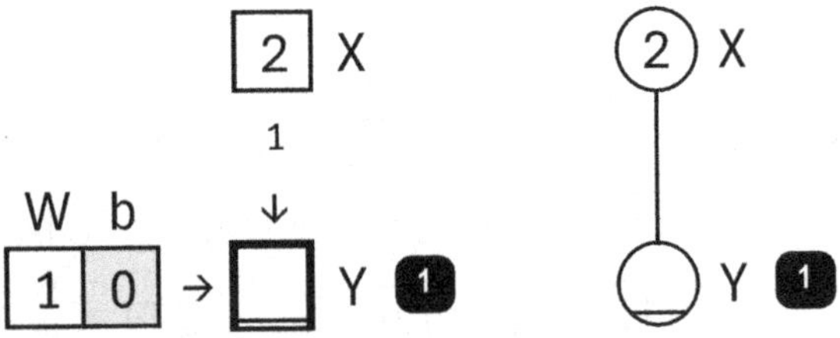

Linear
& ReLU

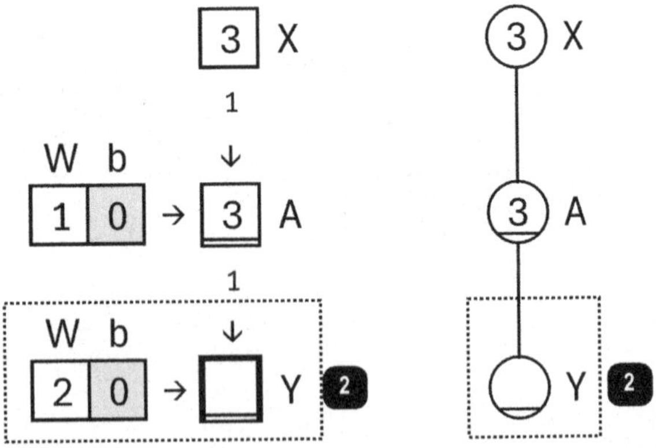

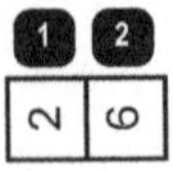

Linear
& ReLU

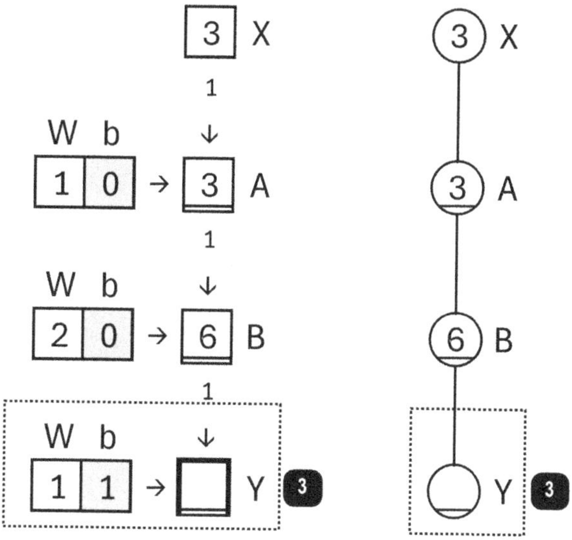

Linear
& ReLU

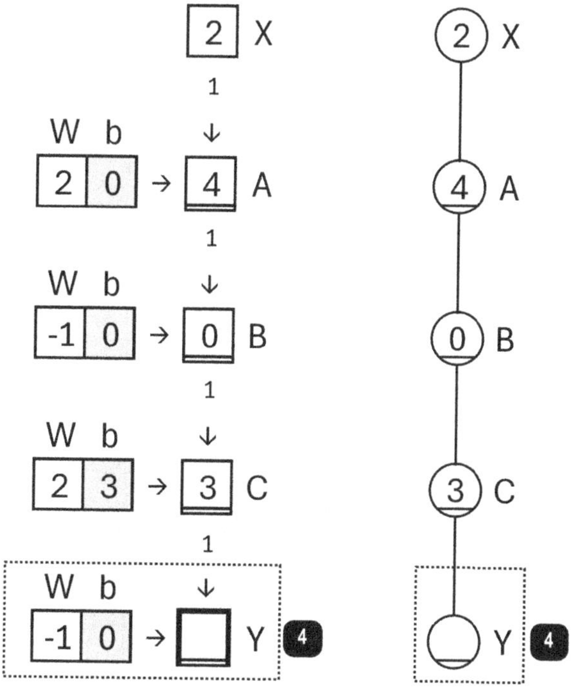

3	4
7	0

Linear
& ReLU

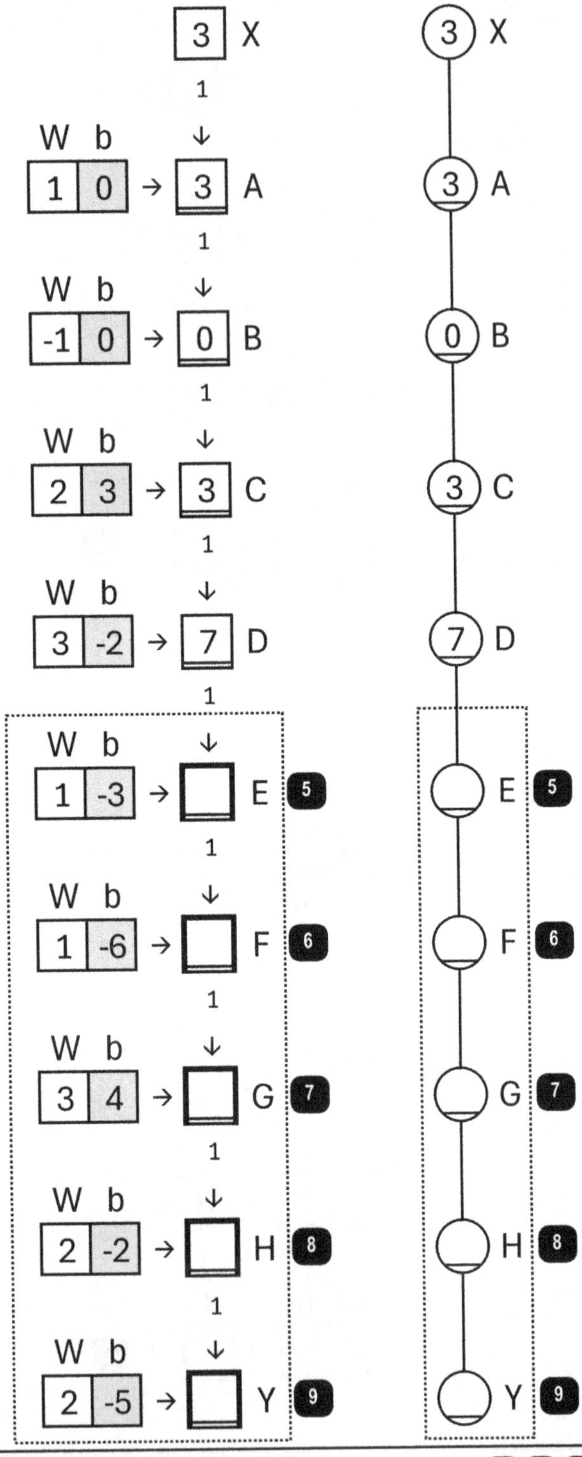

5	6	7	8	9
4	0	4	6	7

Linear
& ReLU

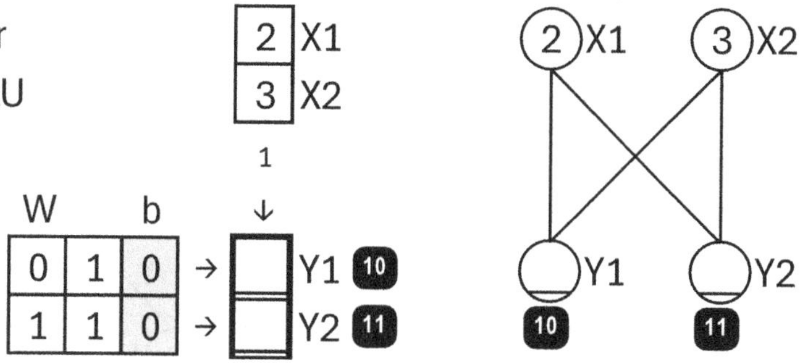

Linear
& ReLU

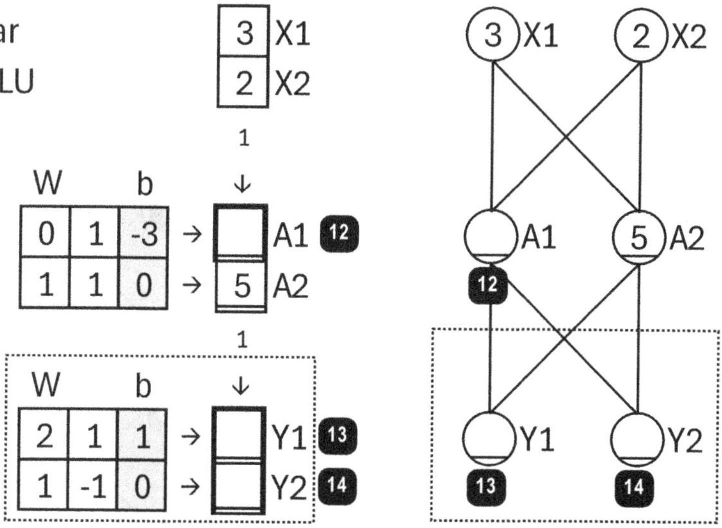

10	11	12	13	14
3	5	0	6	0

Linear
& ReLU

X1	3
X2	2

↓ 1

W		b	→	
0	1	1	→	3 A1
1	0	0	→	3 A2

↓ 1

W		b	→	
1	-1	1	→	1 B1
-1	1	2	→	2 B2

↓ 1

W		b	→	
0	1	-1	→	C1 15
1	1	0	→	C2 16

↓ 1

W		b	→	
0	1	0	→	D1 17
1	0	2	→	D2 18

↓ 1

W		b	→	
1	0	-3	→	E1 19
0	0	1	→	E2 20

↓ 1

W		b	→	
2	0	1	→	Y1 21
1	-1	0	→	Y2 22

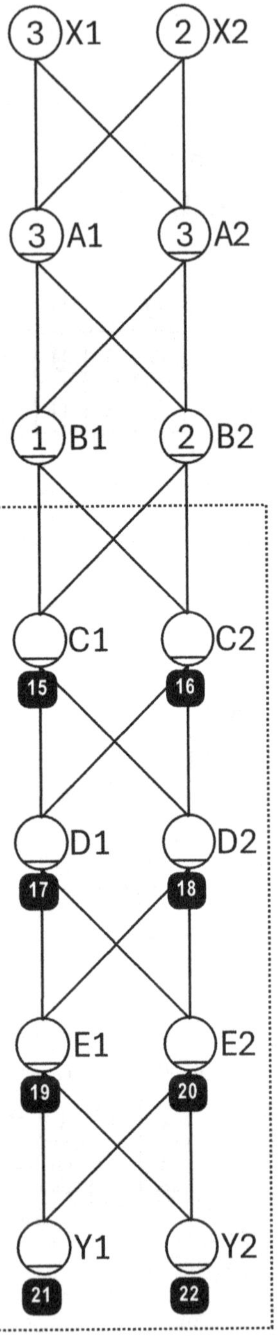

15	16	17	18	19	20	21	22
1	3	3	3	0	1	1	0

Linear
& ReLU

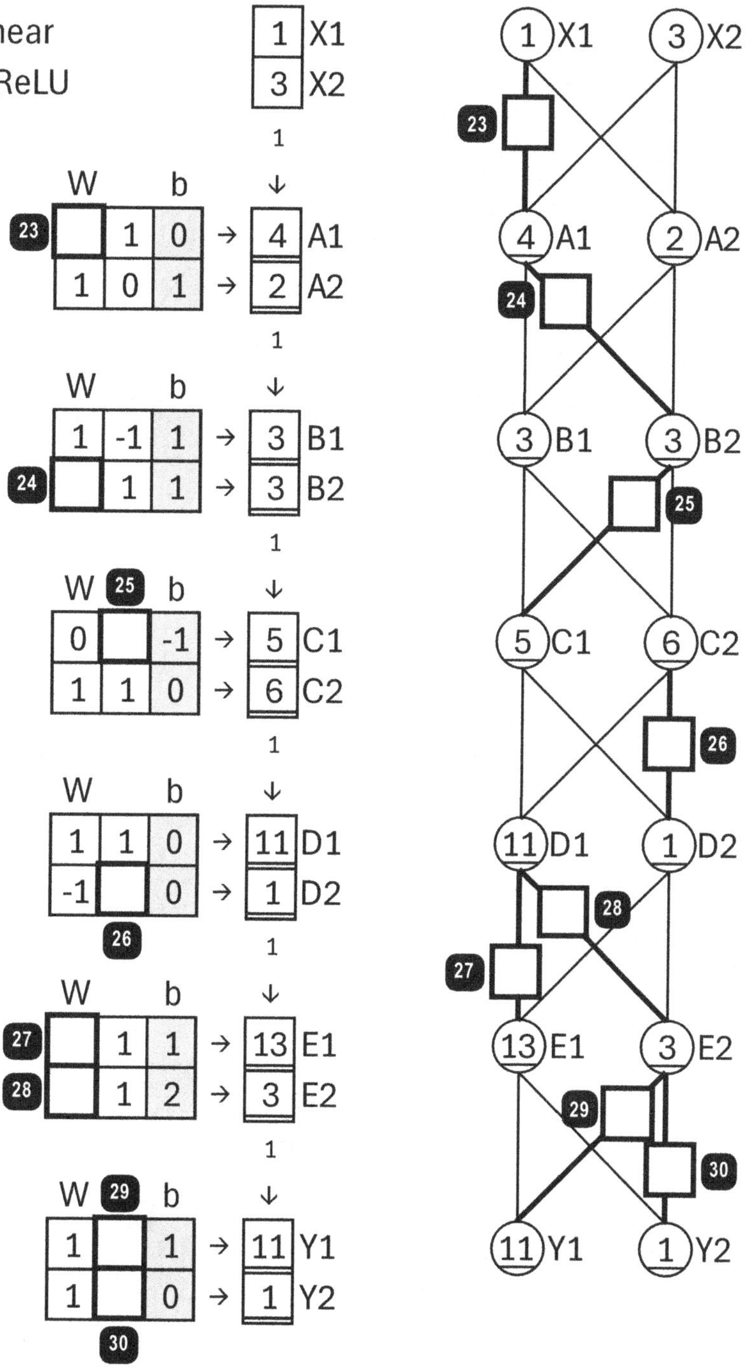

23	24	25	26	27	28	29	30
1	0	2	1	1	0	-1	-4

Linear
& ReLU

3	X1
1	X2
2	X3

1
↓

W			b			
0	1	1	0	→	3	Y1
0	0	1	3	→		Y2 31
1	1	0	0	→		Y3 32

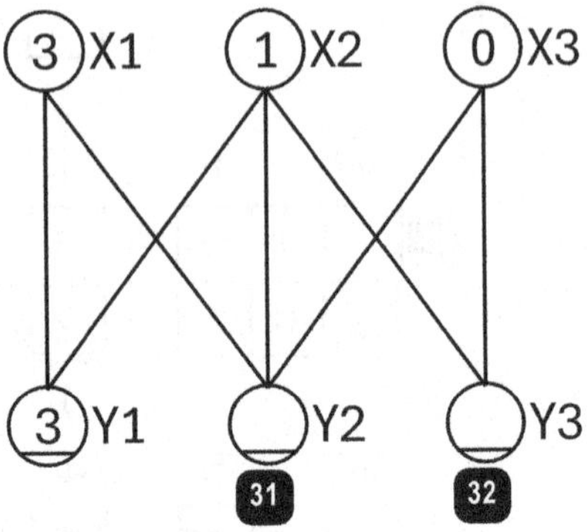

Linear
& ReLU

3	X1
1	X2
2	X3

1
↓

W			b			
1	-1	0	0	→	2	A1
0	0	-1	0	→	0	A2
0	1	0	1	→	2	A3

1
↓

W			b			
0	1	1	1	→	3	Y1
0	0	0	-5	→		Y2 33
1	0	2	0	→		Y3 34

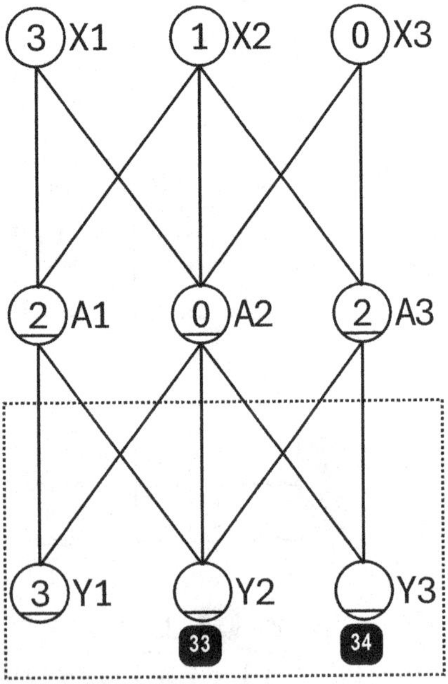

31	32	33	34
5	4	0	6

Linear
& ReLU

3	X1
1	X2
2	X3

1
↓

W			b		
1	-1	0	0	→ 2	A1
0	0	-1	0	→ 0	A2
0	1	0	1	→ 2	A3

1
↓

W			b		
0	1	1	0	→ 2	B1
1	0	-1	2	→ 2	B2
-1	0	3	1	→ 5	B3

1
↓

W			b		
1	1	0	2	→ 6	C1
0	1	1	1	→	C2 35
1	0	0	-5	→	C3 36

1
↓

W			b		
-1	1	0	0	→ 2	D1
1	-1	0	0	→	D2 37
0	1	0	-1	→	D3 38

1
↓

W			b		
1	1	1	0	→ 9	Y1
0	0	2	-9	→	Y2 39
-1	-1	-1	3	→	Y3 40

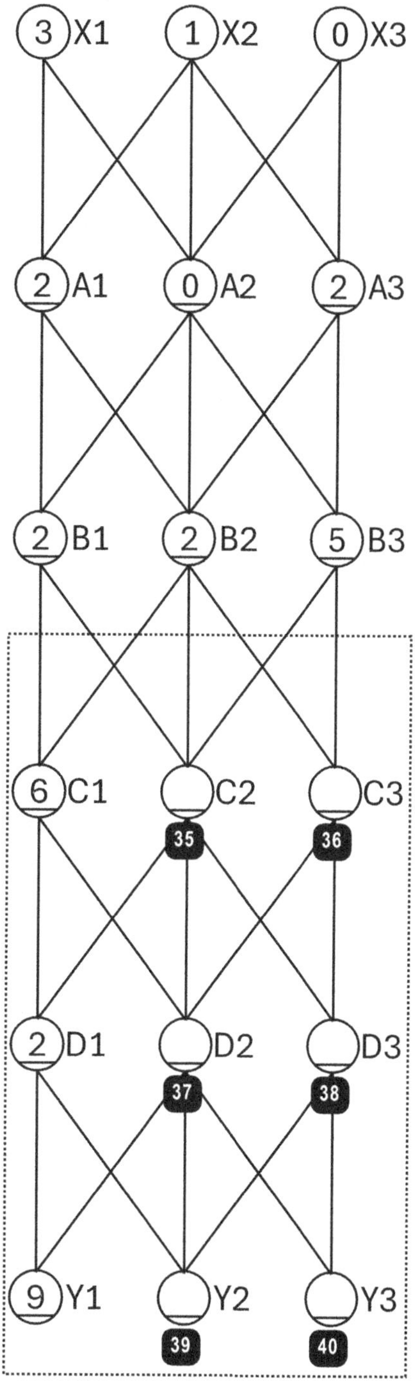

35	36	37	38	39	40
8	0	0	7	5	0

Linear
& ReLU

3	X1
1	X2
2	X3

1 ↓

W			b			
1	0	1	-1	→		A1 (41)
0	1	-1	-2	→	0	A2
1	-1	0	-3	→		A3 (42)

1 ↓

W			b			
0	1	1	0	→	0	B1
1	0	-1	2	→		B2 (43)
-1	0	3	1	→		B3 (44)

1 ↓

W			b			
0	0	0	2	→		C1 (45)
0	1	1	-2	→		C2 (46)
1	1	1	1	→	7	C3

1 ↓

W			b			
(47)	1	0	0	→	2	D1
1	-1	0	3	→	1	D2
0	1	(48)	-1	→	3	D3

1 ↓

W			b			
1	2	-1	-4	→	0	Y1
(49)	0	1	0	→	5	Y2
1	(50)	1	3	→	8	Y3

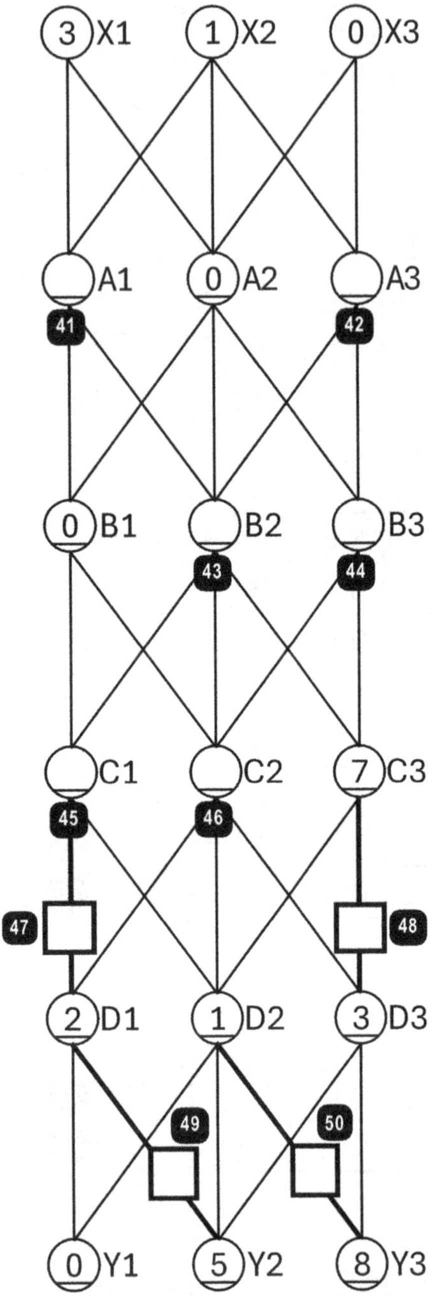

41	42	43	44	45	46	47	48	49	50
4	0	6	0	2	4	-1	0	1	0

Chapter 10

Wide

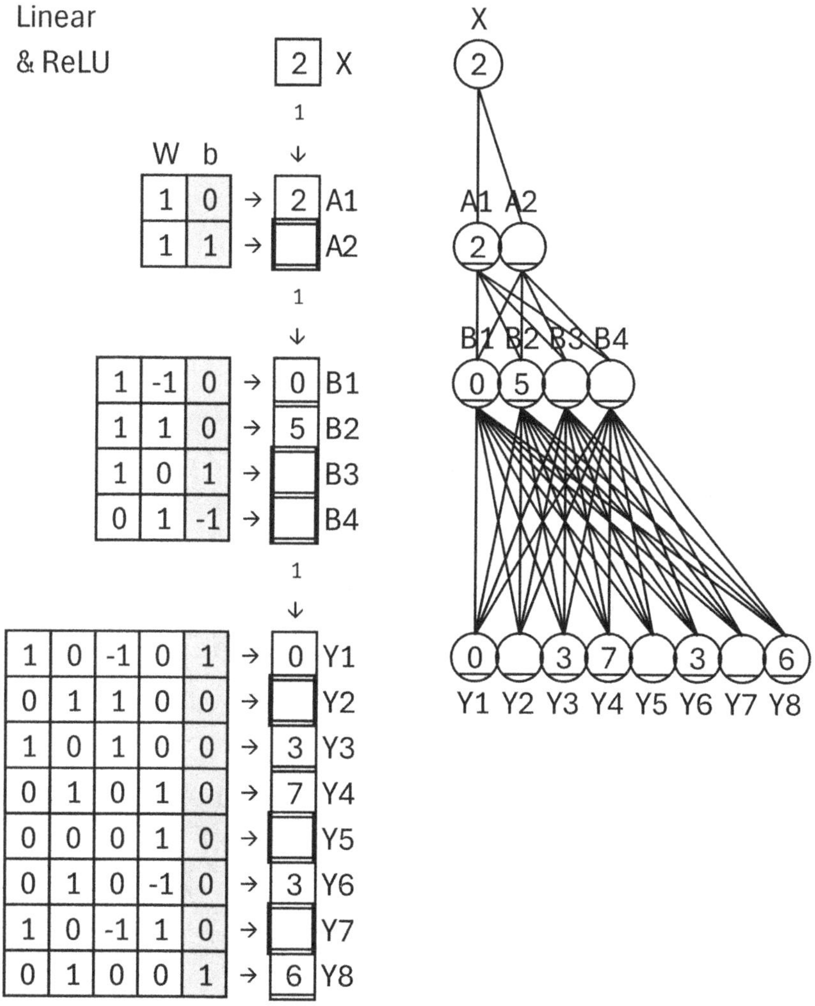

Linear

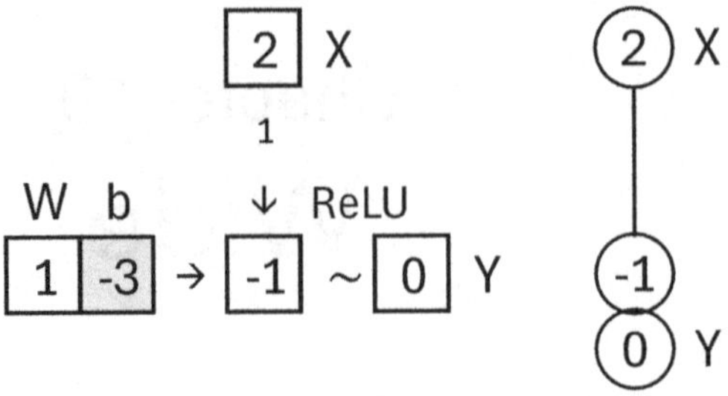

Linear
& ReLU

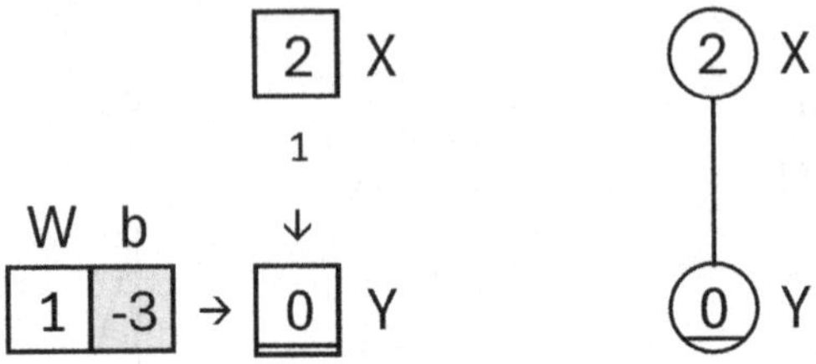

Linear
& ReLU

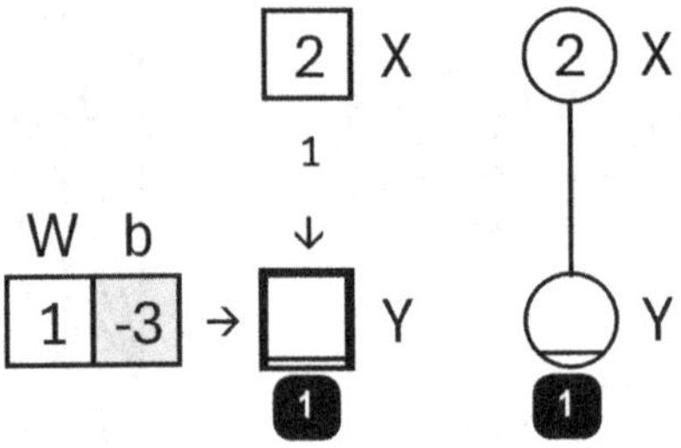

Linear
& ReLU

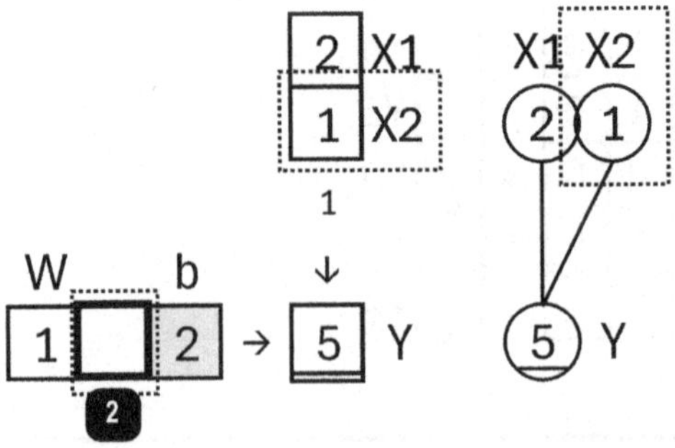

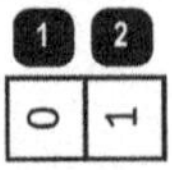

Linear
& ReLU

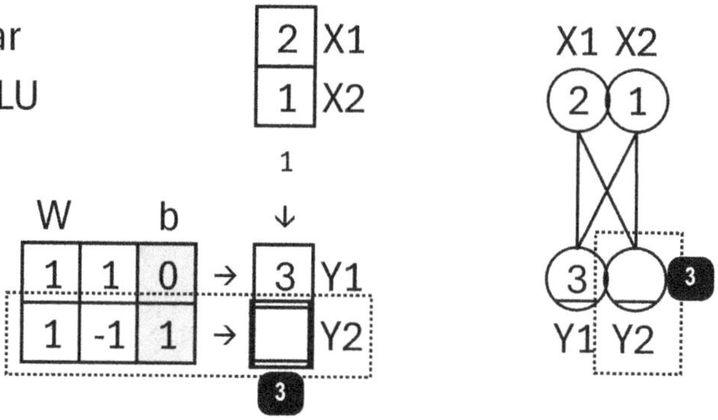

Linear
& ReLU

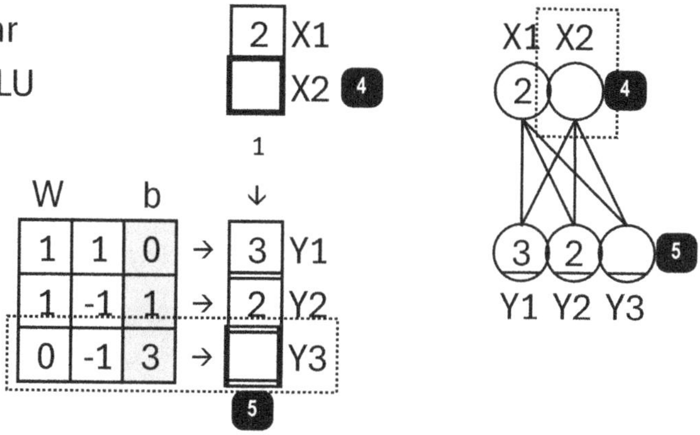

Linear
& ReLU

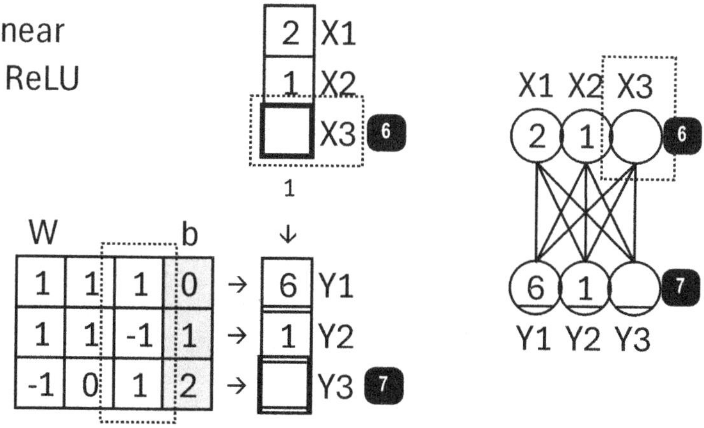

3	4	5	6	7
2	1	2	3	3

Linear
& ReLU

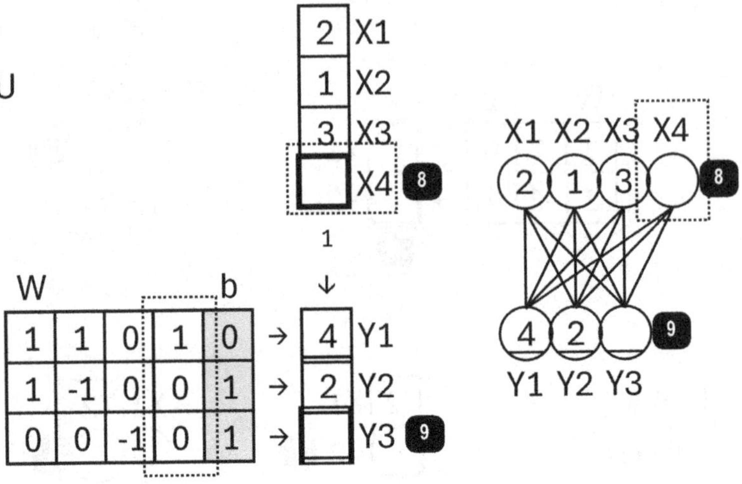

Linear
& ReLU

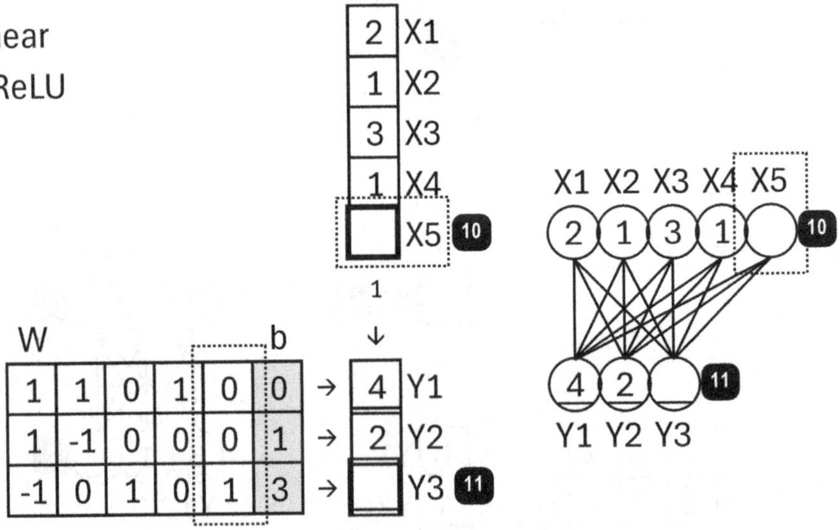

8	9	10	11
1	0	2	6

Linear
& ReLU

2	X1
1	X2
3	X3
1	X4
2	X5
12	X6
1	

W						b			
1	1	0	1	0	0	0	→	4	Y1
1	-1	0	0	0	0	1	→	2	Y2
0	0	1	0	1	-1	1	→		Y3 13

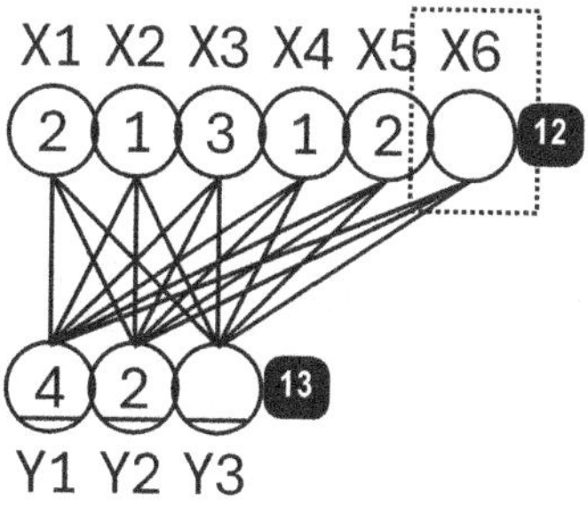

Linear
& ReLU

2	X1
1	X2
3	X3
1	X4
2	X5
1	X6
14	X7
1	

W							b			
1	1	0	1	0	0	0	0	→	4	Y1
1	-1	0	0	0	0	2	0	→	3	Y2
-1	0	1	0	1	-1	0	1	→		Y3 15

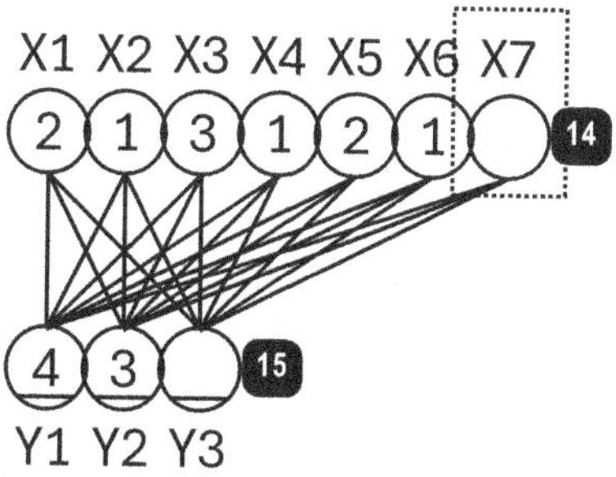

12	13	14	15
1	5	1	3

Linear
& ReLU

X	
2	X1
1	X2
3	X3
1	X4
2	X5
1	X6
1	X7

1
↓

W						16	b			
0	1	1	0	0	0		0	→	4	Y1
1	-1	0	0	1	0	2	0	→	5	Y2
-1	0	1	0	1	-1	0	1	→	3	Y3
0	2	0	0	0	0	0	-3	→		Y4 17

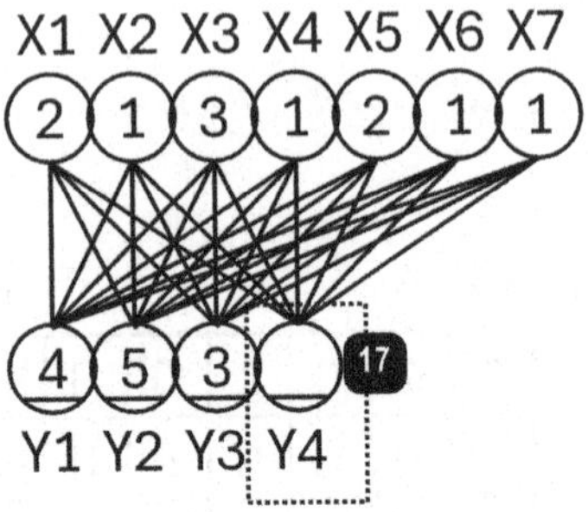

Linear
& ReLU

X	
2	X1
1	X2
3	X3
1	X4
2	X5
1	X6
1	X7

1
↓

W						18	b			
1	1	0	0	0	0		0	→	4	Y1
1	-1	0	0	0	0	2	0	→	3	Y2
-1	0	1	0	1	-1	0	1	→	3	Y3
0	2	0	0	0	0	1	-2	→		Y4 19
0	0	0	0	-1	0	1	0	→		Y5 20

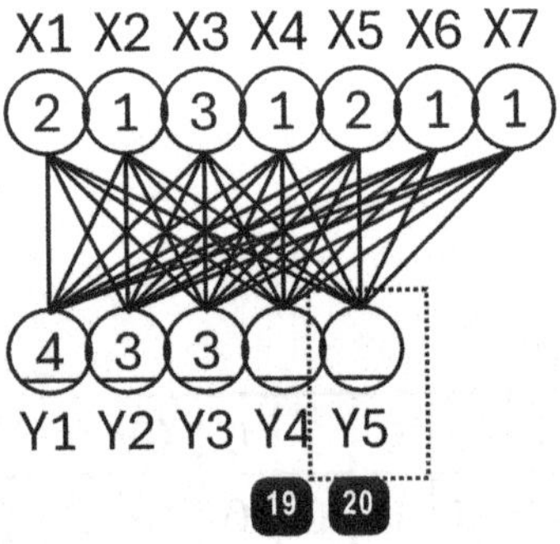

16	17	18	19	20
0	0	1	1	0

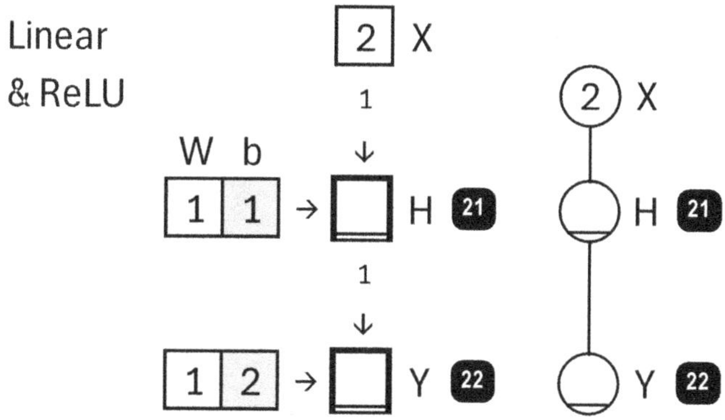

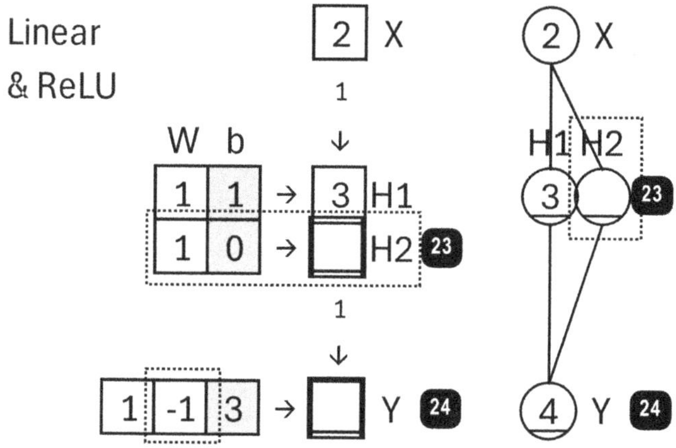

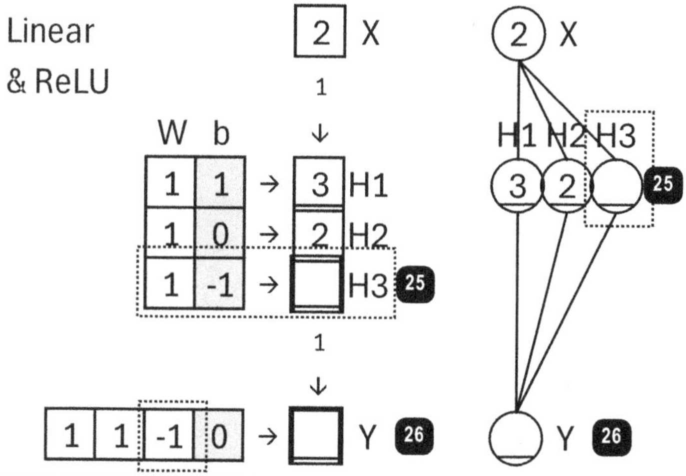

21	22	23	24	25	26
3	5	2	4	1	4

Linear
& ReLU

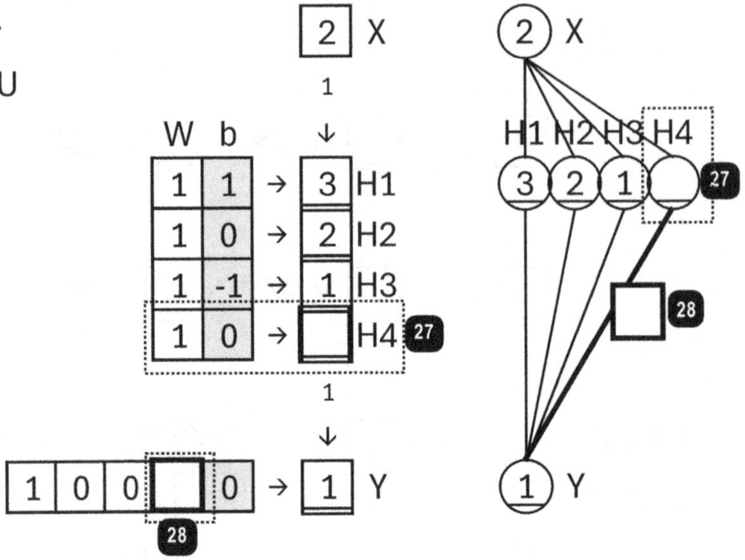

Linear
& ReLU

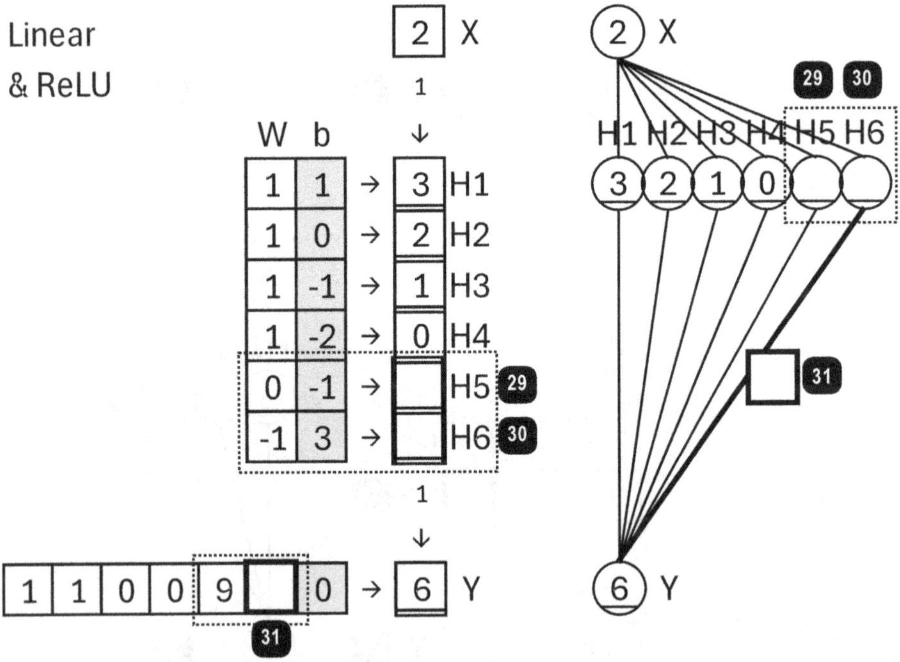

27	28	29	30	31
2	-1	0	1	1

Linear
& ReLU

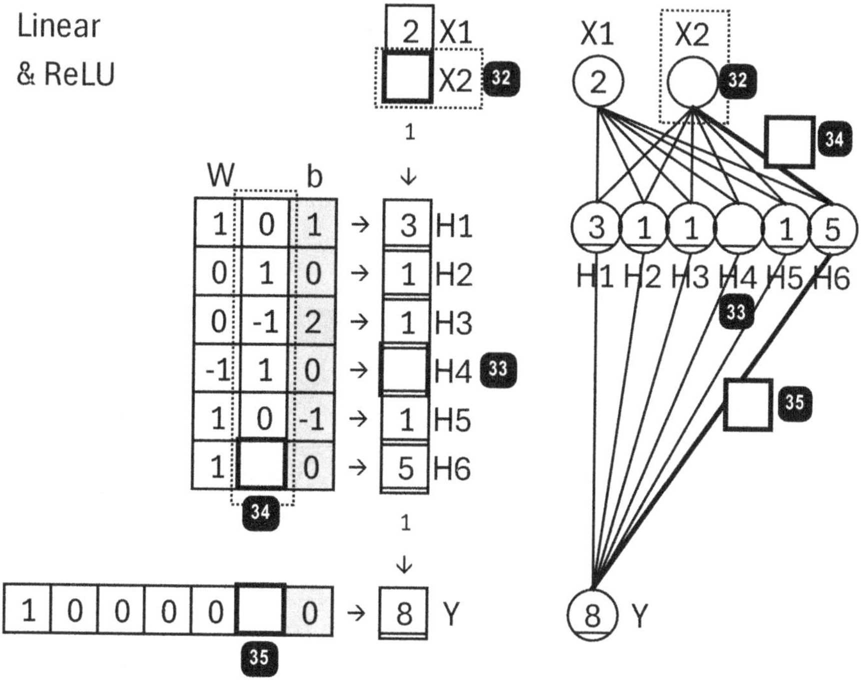

Linear
& ReLU

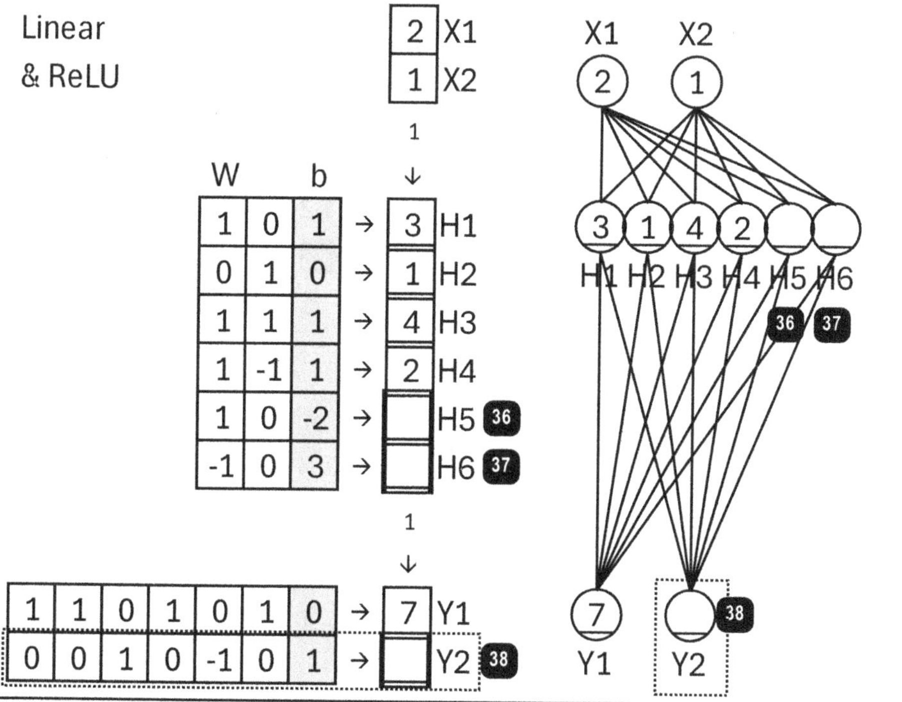

32	33	34	35	36	37	38
1	0	3	1	0	1	5

Linear
& ReLU

			2	X
			1 ↓	
W	b			
1	0	→	2	A1
1	1	→		A2 (39)
			1 ↓	

1	-1	0	→	0	B1
1	1	0	→	5	B2
1	0	1	→		B3 (40)
0	1	-1	→		B4 (41)
				1 ↓	

1	0	-1	0	1	→	0	Y1
0	1	1	0	0	→		Y2 (42)
1	0	1	0	0	→	3	Y3
0	1	0	1	0	→	7	Y4
0	0	0	1	0	→		Y5 (43)
0	1	0	-1	0	→	3	Y6
1	0	-1	1	0	→		Y7 (44)
0	1	0	0	1	→	6	Y8

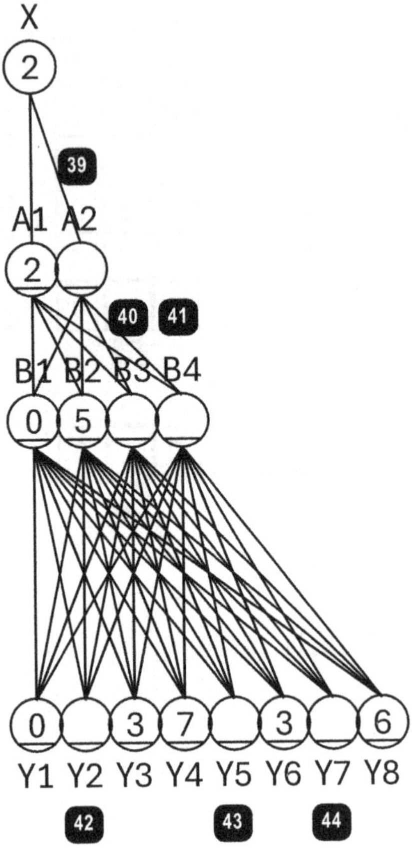

39	40	41	42	43	44
3	3	2	8	2	0

Linear
& ReLU

2	X1
0	X2
-1	X3
4	X4
1	X5
2	X6
3	X7

1
↓

1	0	0	0	-1	0	0	1	→		A1 45
0	0	0	1	0	1	0	0	→	6	A2
0	1	0	0	0	0	0	-1	→		A3 46
0	0	1	1	0	0	0	0	→	3	A4
0	0	0	0	1	1	0	-1	→		A5 47

1
↓

0	1	1	1	0	0	→		B1 48
0	0	0	1	1	1	→	6	B2
1	1	0	0	0	0	→		B3 49

1
↓

-1	1	-1	0	→		Y 50

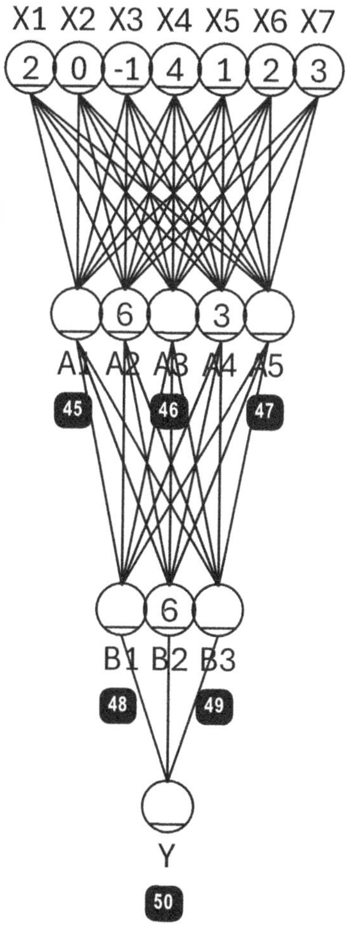

45	46	47	48	49	50
2	0	2	9	8	0

Chapter 11
Softmax

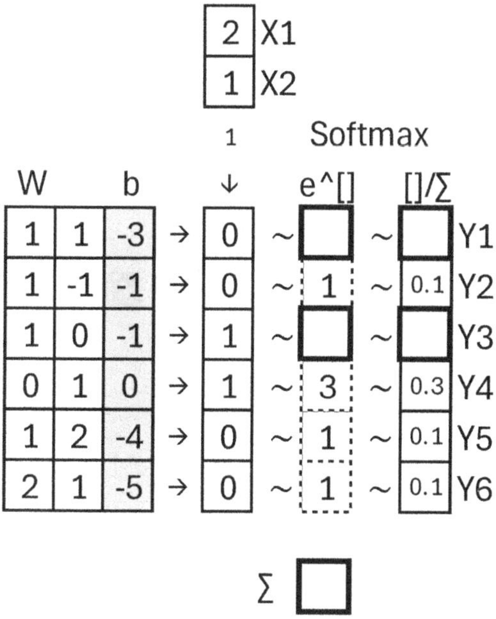

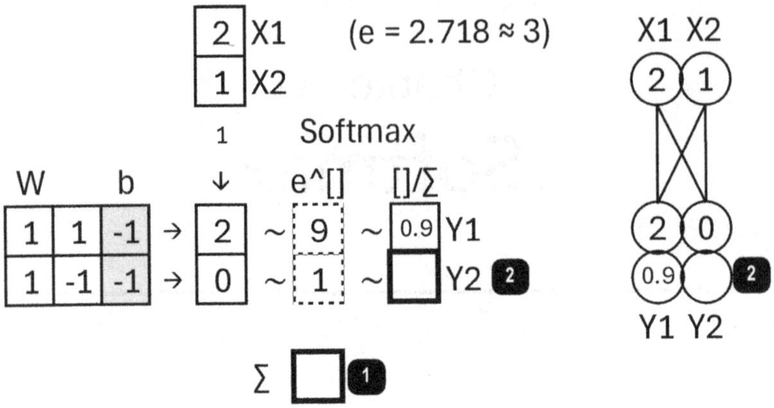

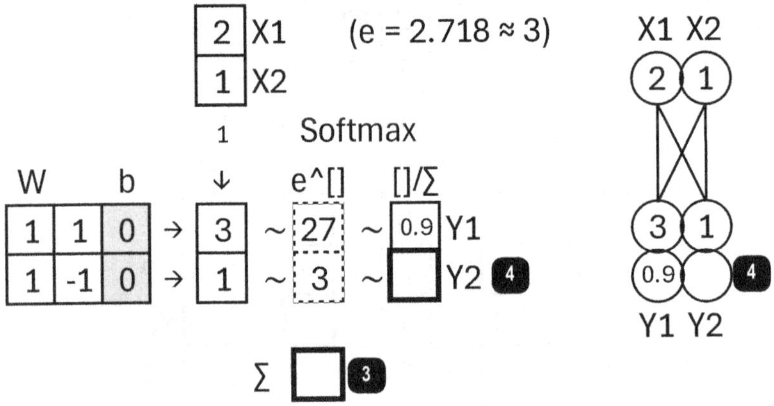

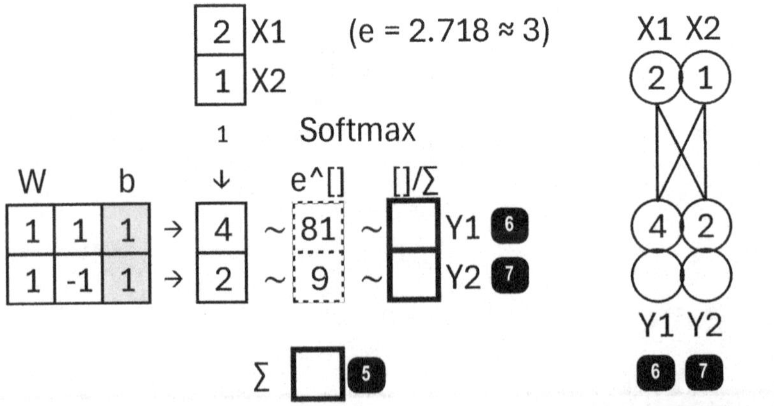

1	2	3	4	5	6	7
10	0.1	30	0.1	90	0.9	0.1

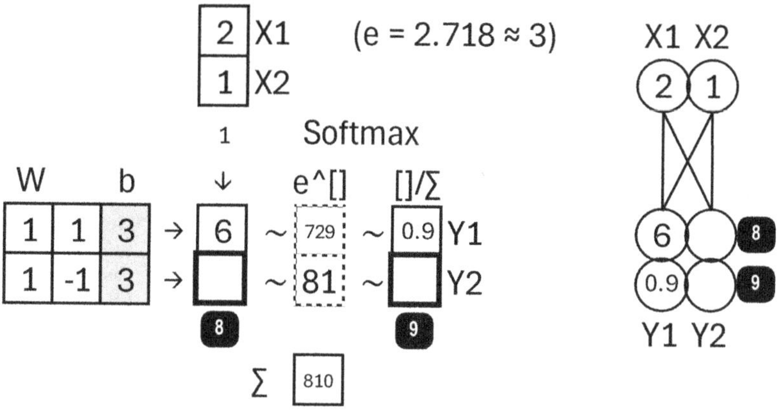

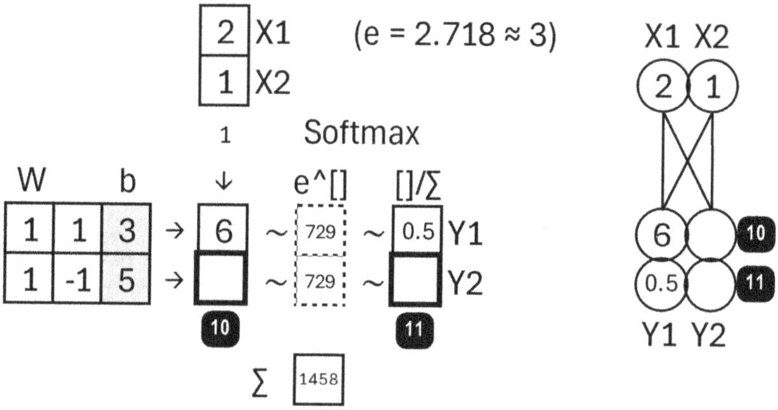

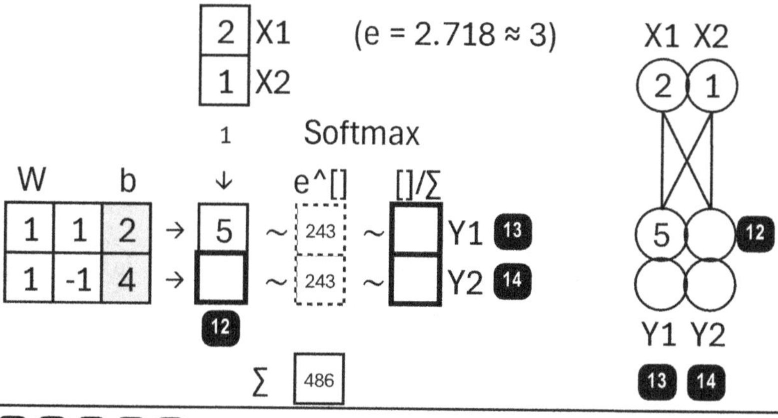

8	9	10	11	12	13	14
4	0.1	6	0.5	5	0.5	0.5

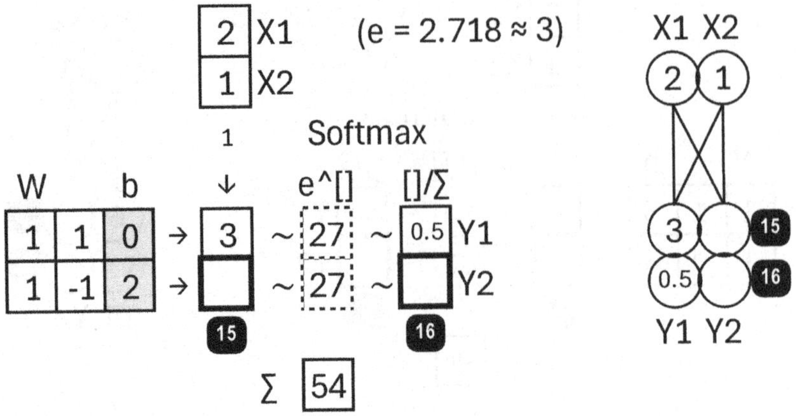

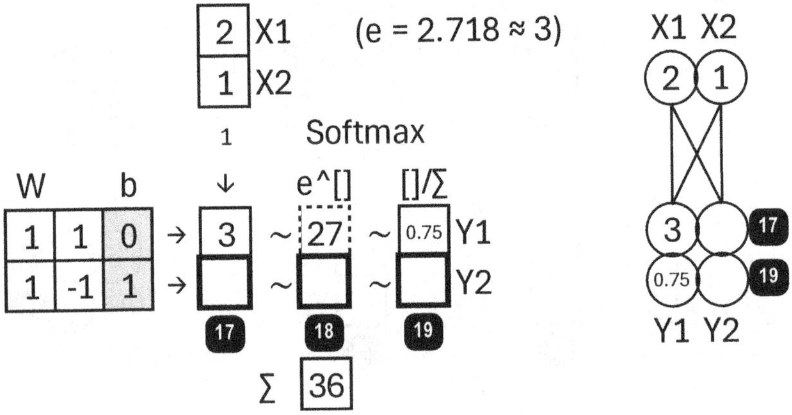

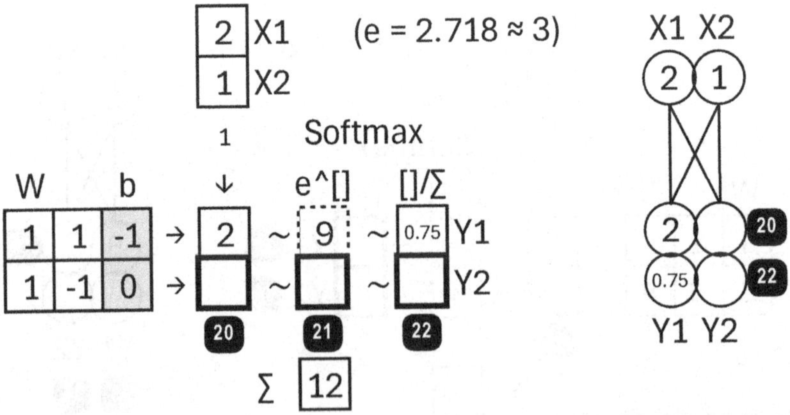

15	16	17	18	19	20	21	22
3	0.5	2	9	0.25	1	3	0.25

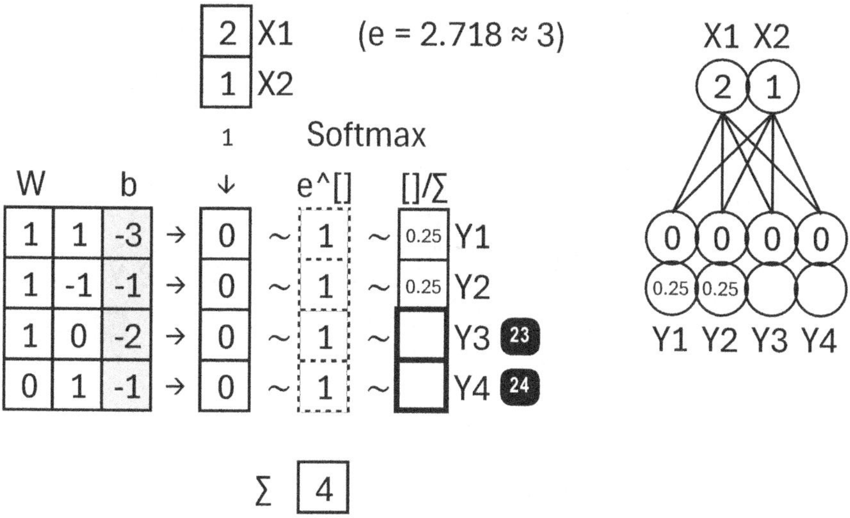

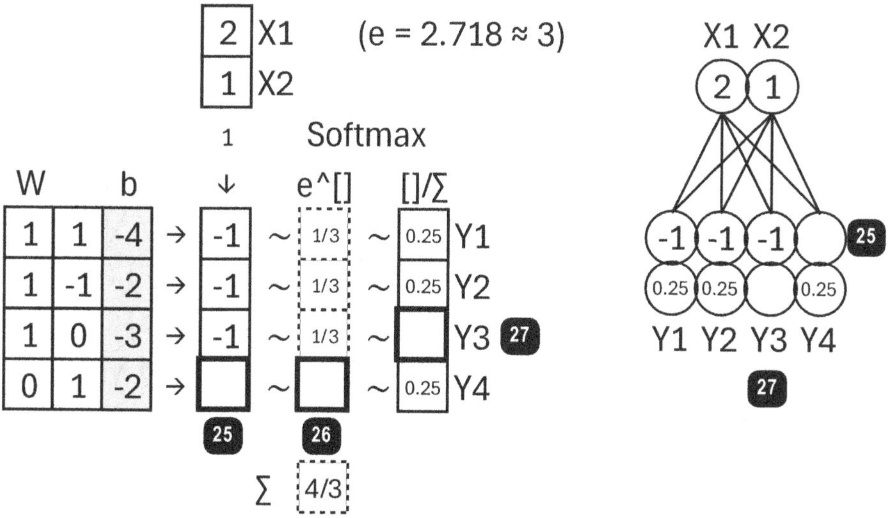

23	24	25	26	27
0.25	0.25	-1	1/3	0.25

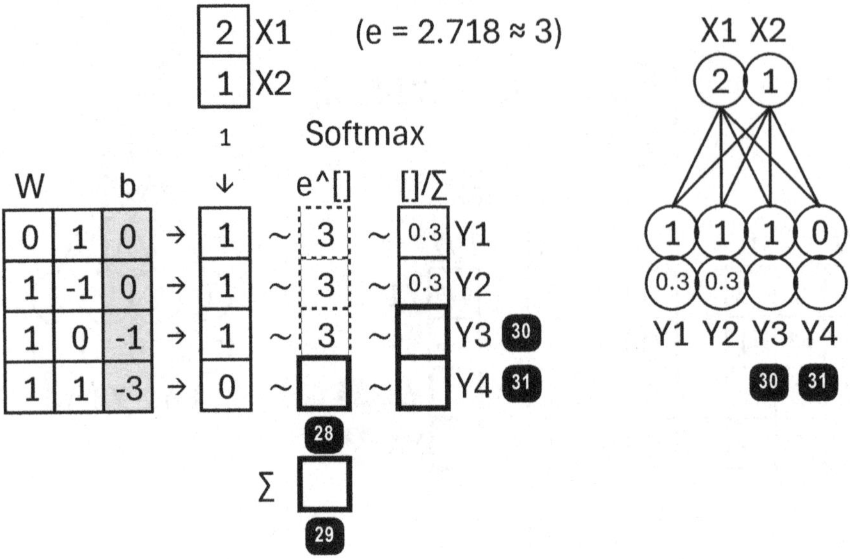

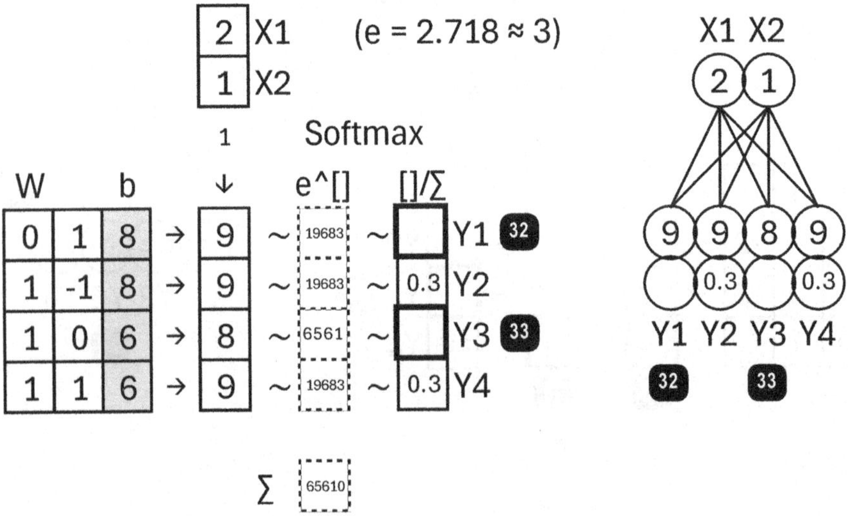

28	29	30	31	32	33
1	10	0.3	0.1	0.3	0.1

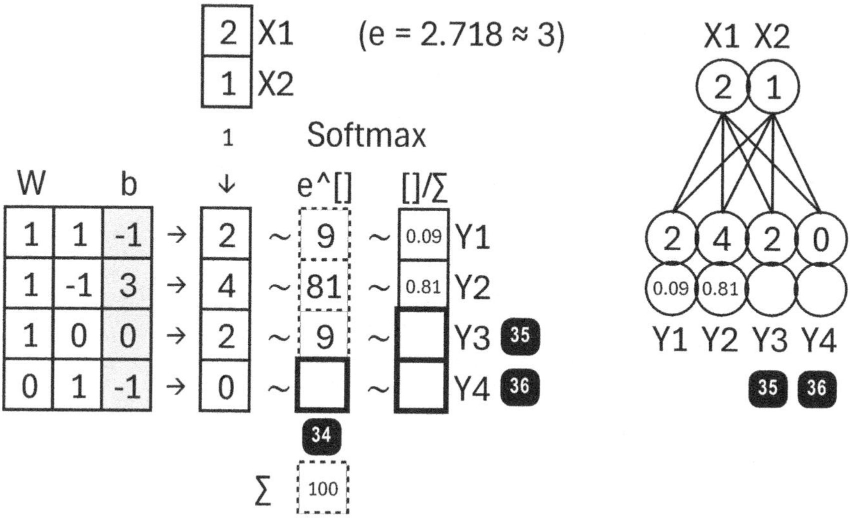

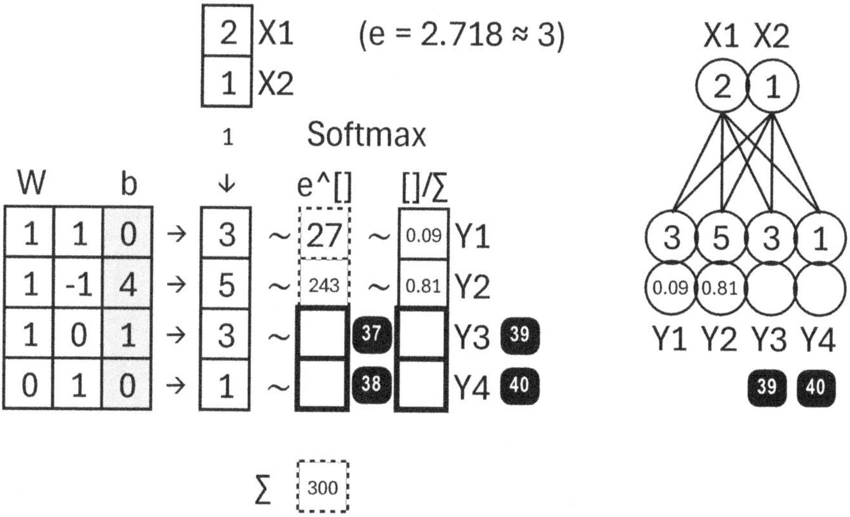

34	35	36	37	38	39	40
1	0.09	0.01	27	3	0.09	0.01

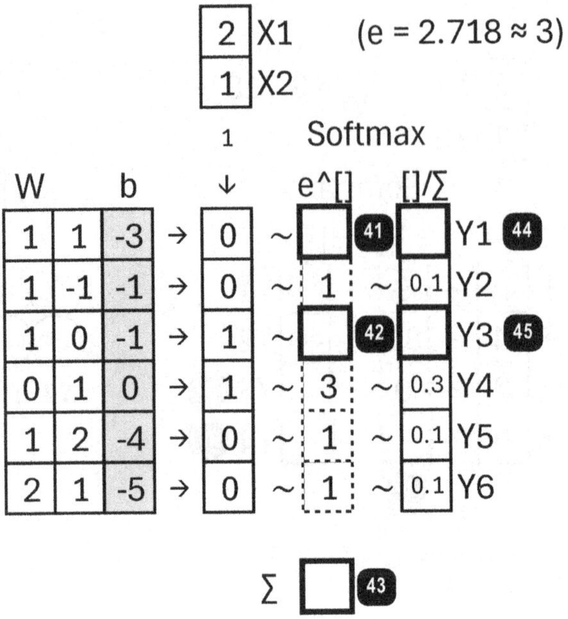

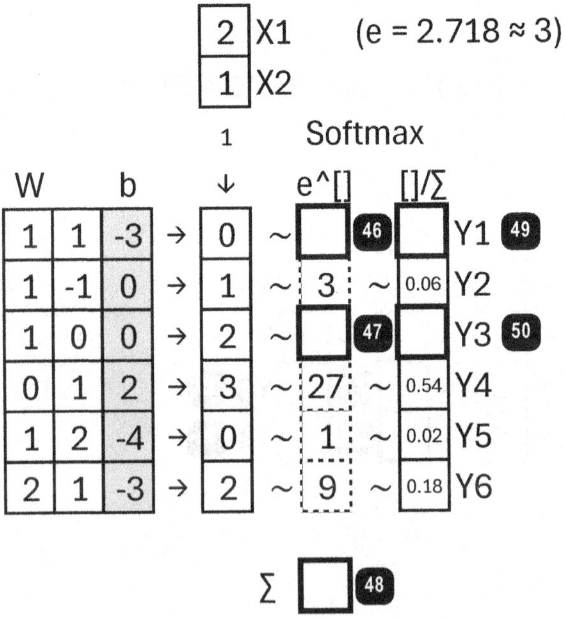

41	42	43	44	45	46	47	48	49	50
1	3	10	0.1	0.3	1	9	50	0.02	0.18

Chapter 12

Gradient

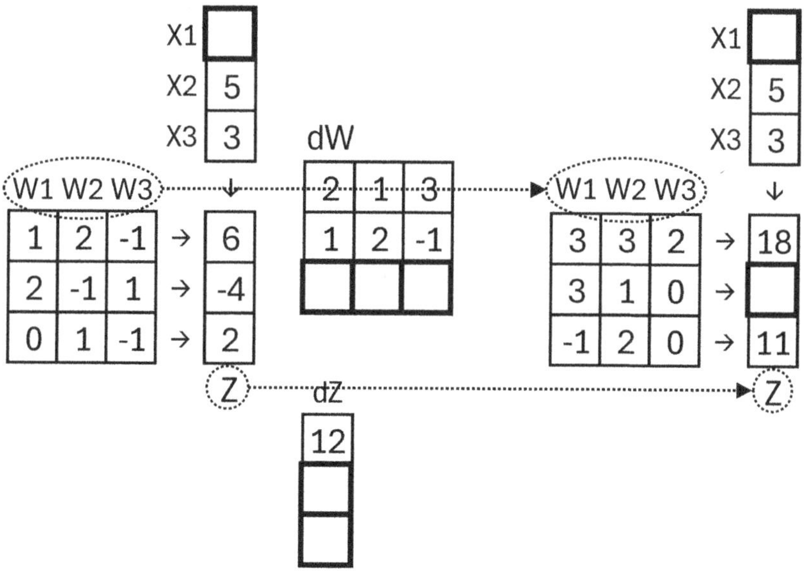

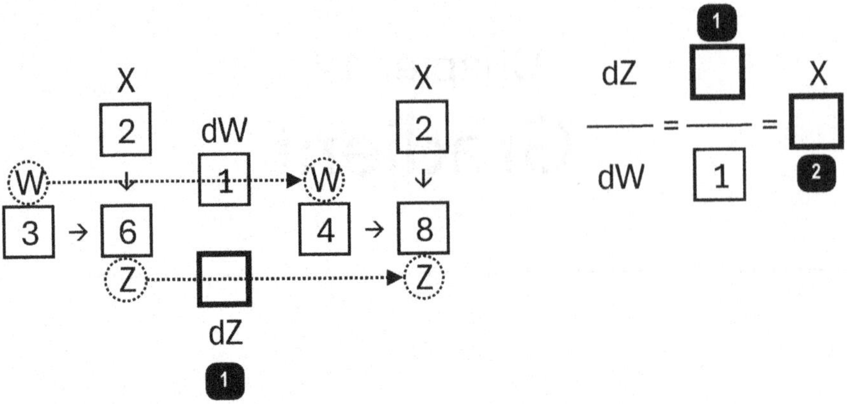

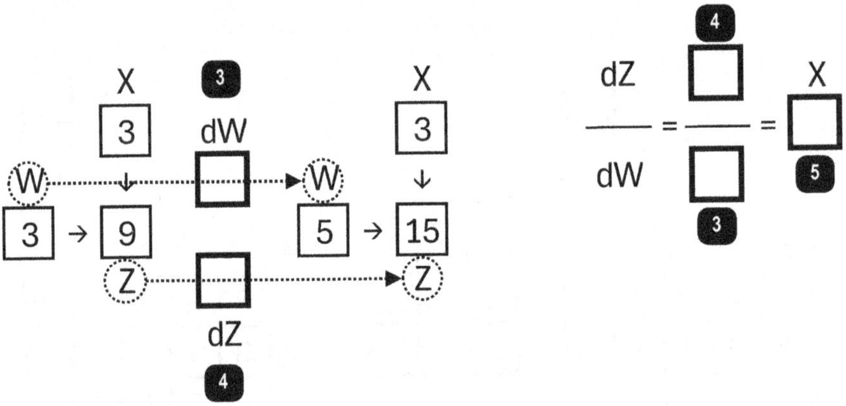

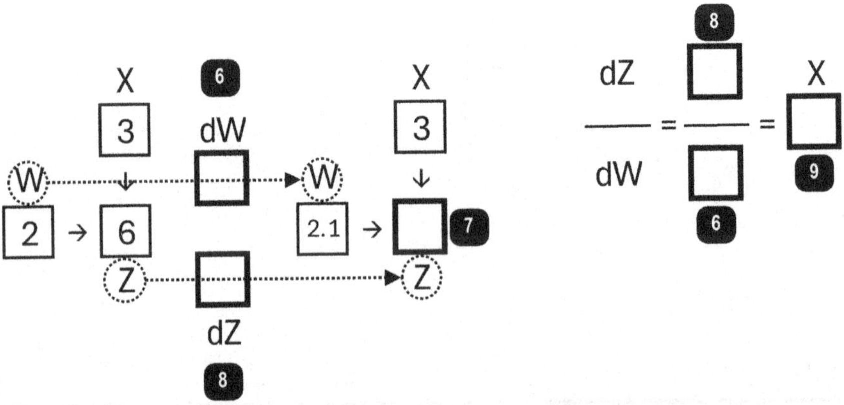

1	2	3	4	5	6	7	8	9
2	2	2	6	3	0.1	6.3	0.3	3

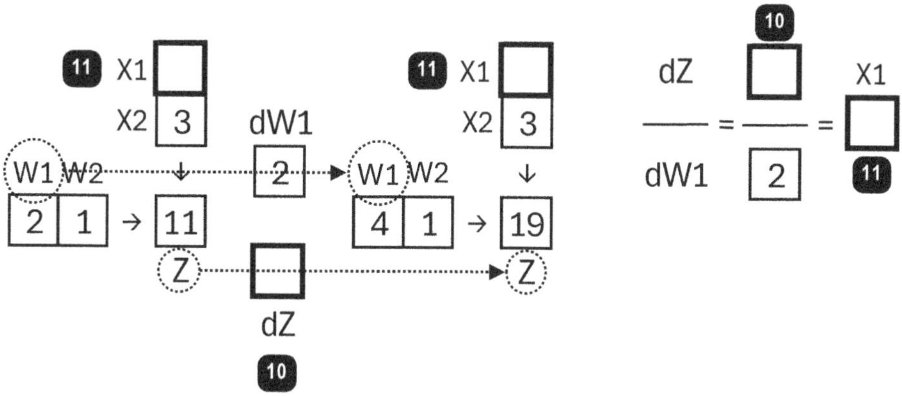

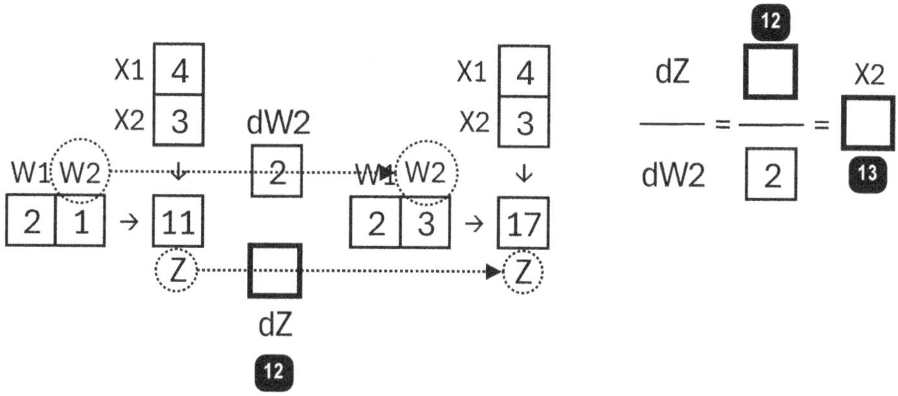

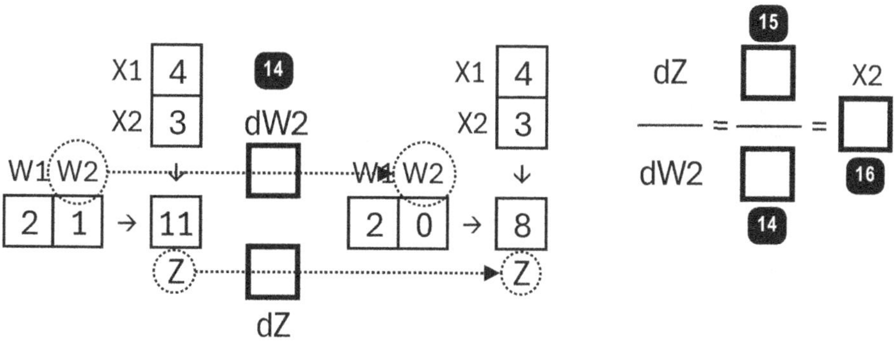

10	11	12	13	14	15	16
8	4	6	3	-1	-3	3

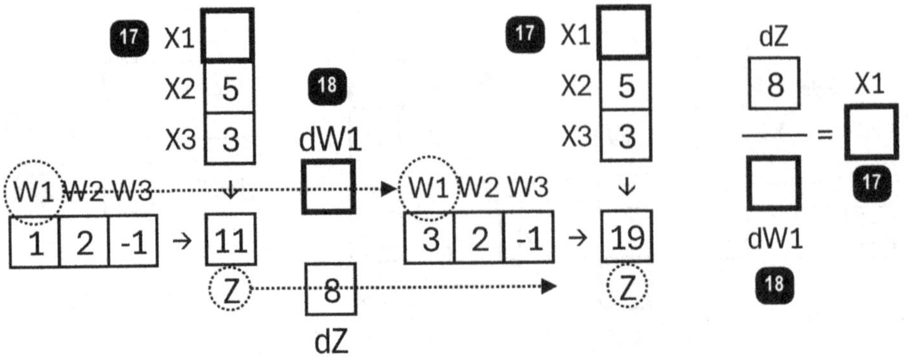

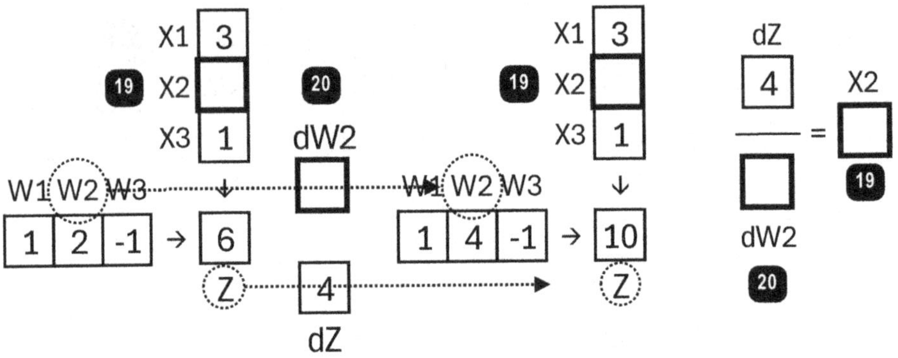

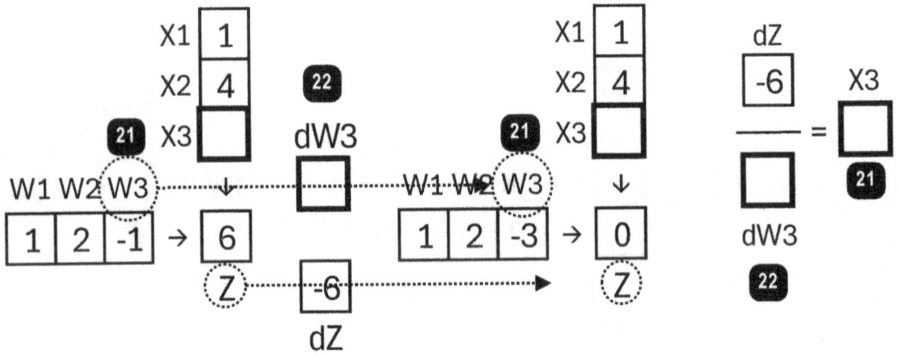

17	18	19	20	21	22
4	2	2	2	3	-2

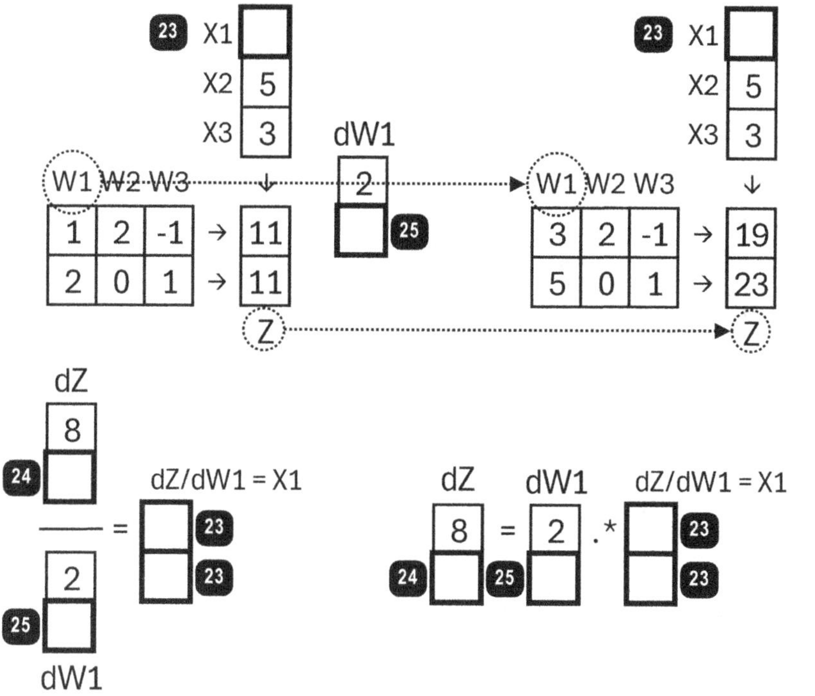

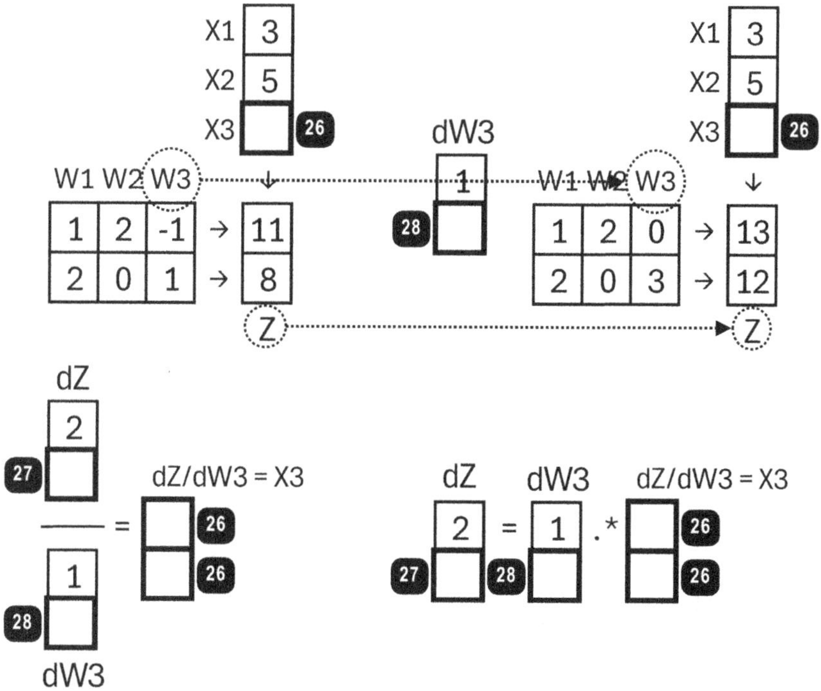

23	24	25	26	27	28
4	12	3	2	4	2

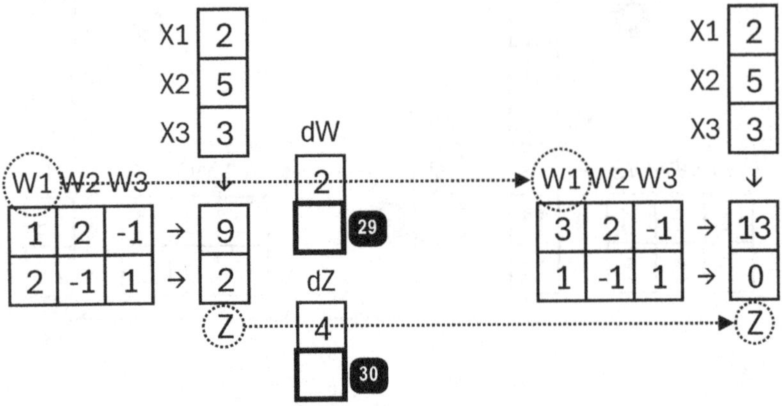

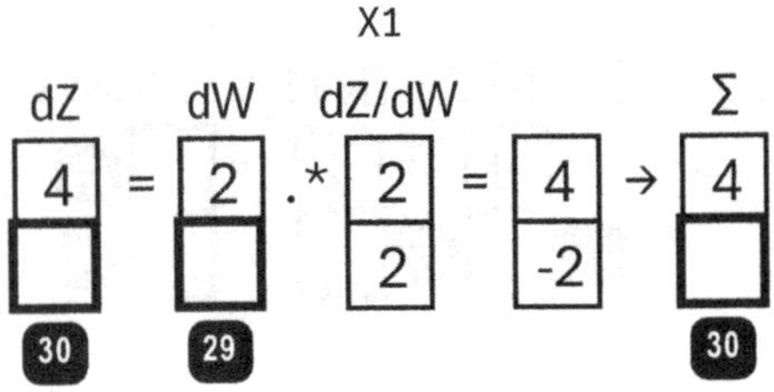

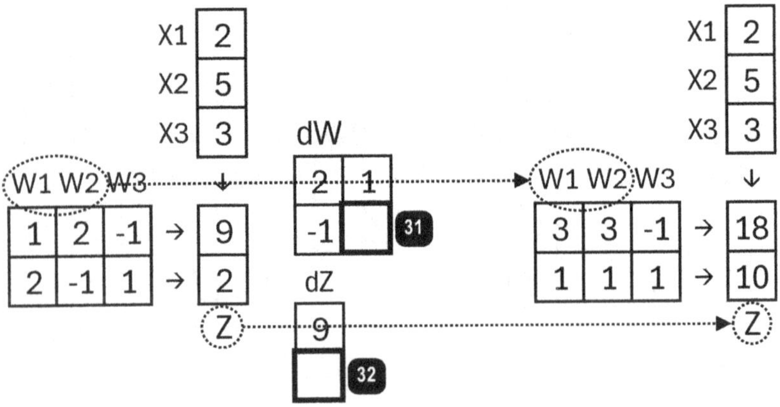

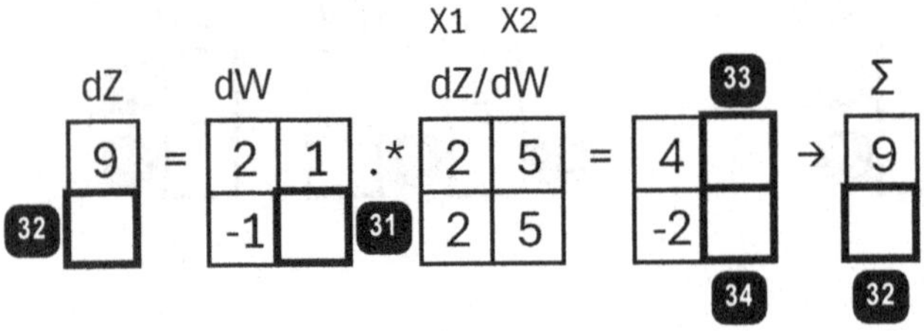

29	30	31	32	33	34
-1	-2	2	8	5	10

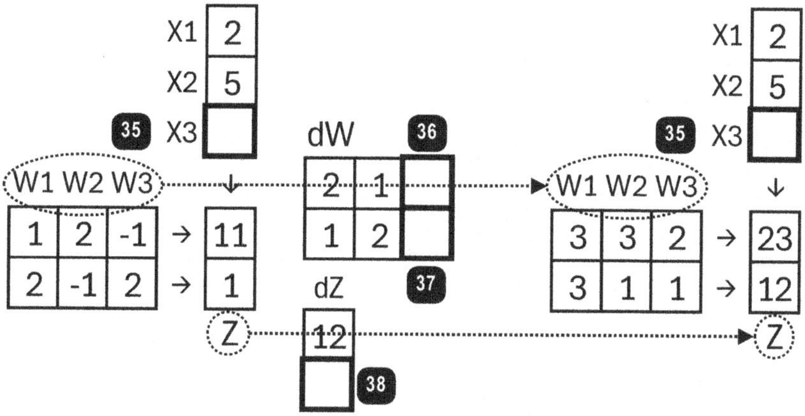

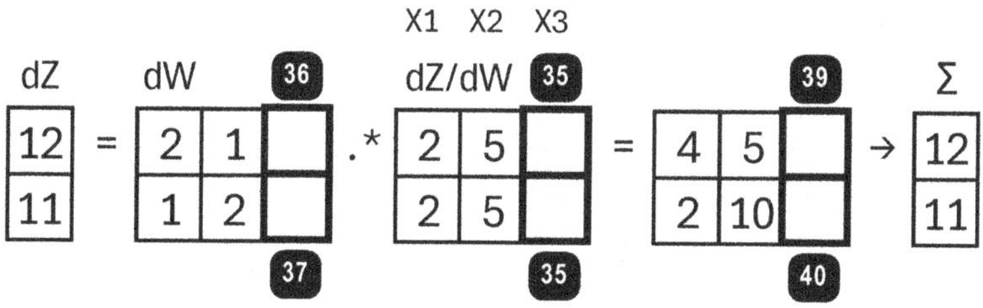

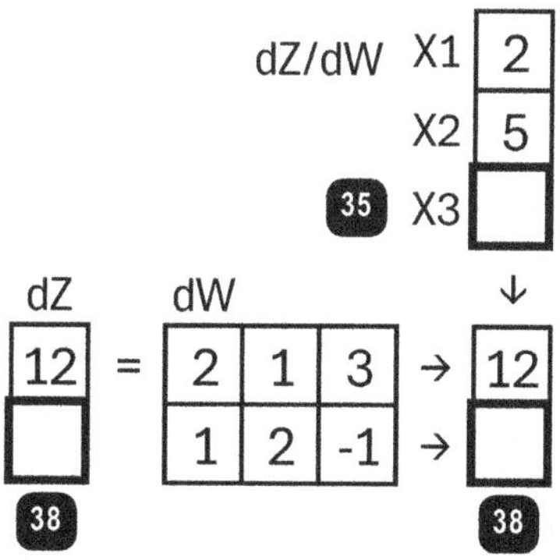

35	36	37	38	39	40
1	3	-1	11	3	-1

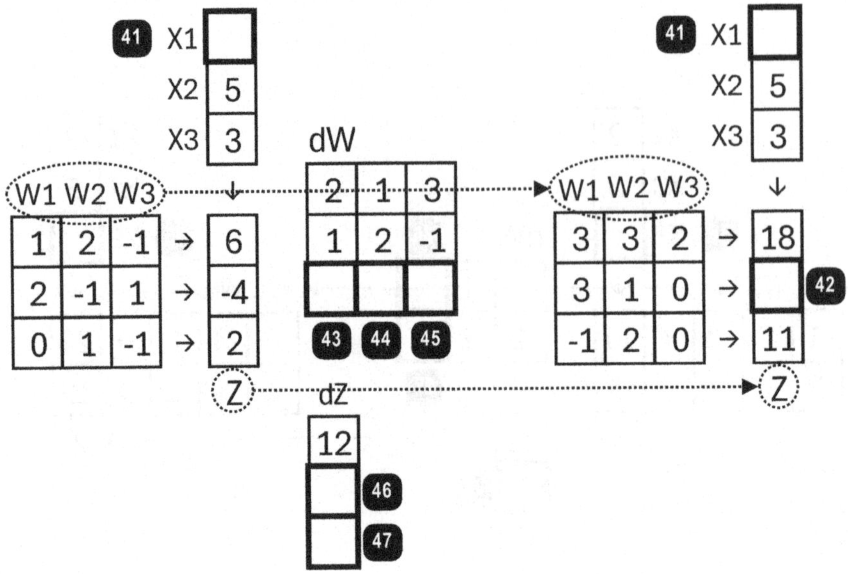

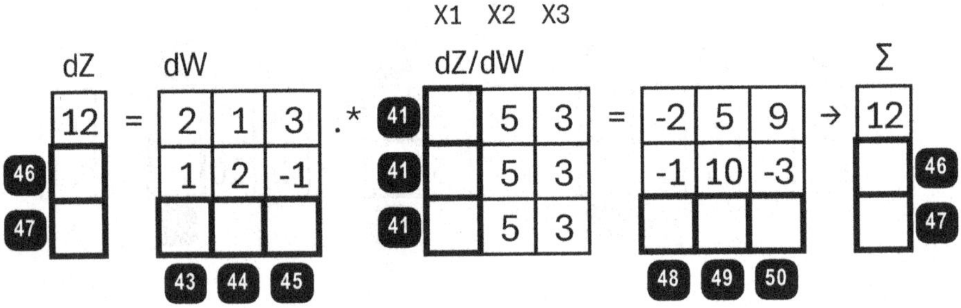

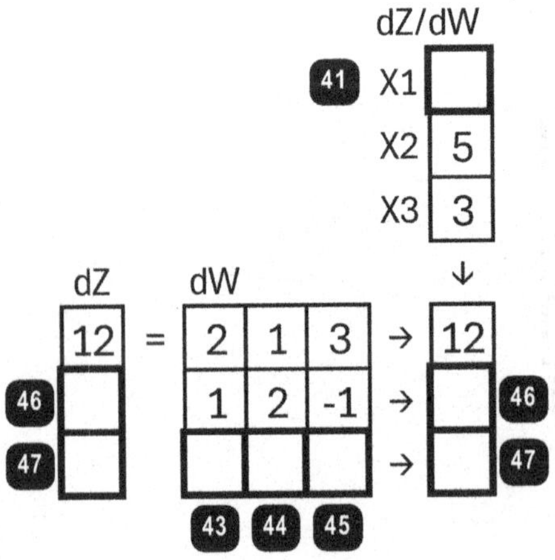

41	42	43	44	45	46	47	48	49	50
-1	2	-1	1	1	6	9	1	5	3

Made in United States
Orlando, FL
04 November 2026

90201635R10066